*Extreme Stars* is a unique book describing the lives of stars from a new perspective. It examines their amazing extremes and results in a refreshing, up-to-date and engaging overview of stellar evolution, suitable for anyone interested in, viewing, or studying stars. Ten chapters, generously illustrated throughout, explain the natures of the brightest, the largest, the hottest, the youngest, and so on, ending with a selection of the strangest stars the Universe has to offer. Taken as a whole, the chapters show how stars develop and die and how each extreme turns into another under the inexorable twin forces of time and gravity.

**James Kaler** is Professor of Astronomy at the University of Illinois at Urbana–Champaign. His research, in which he has published over 100 papers, involves dying stars, specifically the graceful shells and rings of gas ejected in stellar death called planetary nebulae. He has held both Fulbright and Guggenheim Fellowships, and has been awarded medals for his work from the University of Liège, in Belgium, and the University of Mexico.

Long interested in science education and popularization, Dr Kaler has written for a variety of magazines that include *l'Astronomia* (Italy), *Astronomy*, *Sky and Telescope*, *Stardate*, and *Scientific American*, and has written several widely acclaimed books, including *Stars and their Spectra* (Cambridge University Press, 1989, 1997). He lectures widely and appears frequently on regional Illinois television and radio, and recently received the Armand Spitz Memorial Lectureship from the Great Lakes Planetarium Association for his work.

Extreme Stars

# Extreme Stars

## At the Edge of Creation

James B. Kaler
*University of Illinois, Urbana–Champaign*

CAMBRIDGE
UNIVERSITY PRESS

CAMBRIDGE UNIVERSITY PRESS
Cambridge, New York, Melbourne, Madrid, Cape Town, Singapore,
São Paulo, Delhi, Dubai, Tokyo, Mexico City

Cambridge University Press
The Edinburgh Building, Cambridge CB2 8RU, UK

Published in the United States of America by Cambridge University Press, New York

www.cambridge.org
Information on this title: www.cambridge.org/9780521158022

First published 2001
Reprinted 2002
First paperback edition 2010

*A catalogue record for this publication is available from the British Library*

*Library of Congress Cataloguing in Publication Data*

Kaler, James B.
Extreme stars : at the edge of creation / James B. Kaler.
    p. cm.
Includes index.
ISBN 0 521 40262 X
1. Stars. I. Title.
QB801.K23 2001
523.8–dc21                    00–058522

ISBN 978-0-521-40262-0 Hardback
ISBN 978-0-521-15802-2 Paperback

To Marilyn and Len Teitelbaum and Lillis Caulton, with love, and with deepest thanks for always being there.

# Contents

*These plates are available in color for download from
www.cambridge.org/9780521158022

# Tables

# Acknowledgements

This work began as a series of articles that were requested by Professor Corrado Lamberti, editor of the Italian astronomy magazine *l'Astronomia*. They were later published in somewhat different form in the US magazine *Astronomy*, whose editors subsequently commissioned additional articles. These pieces, expanded and completely rewritten, form the basis of the chapters to follow. They are linked to show the flow of stellar evolution from a unique perspective, as viewed from the edges of the HR (Hertzsprung–Russell) diagram.

I would like to express my deepest thanks to Professor Lamberti for first suggesting the project, for suggesting several topics, and for producing the original articles and accompanying graphics with great style. Similar thanks go to the editors of *Astronomy*, who also believed in the viability of the project and who not only provided excellent support and graphics but who also commissioned some fine original art. Hats off in particular (in no order whatsoever) to Richard Berry, Robert Burnham, Alan Dyer, Dave Eicher, Rich Talcott, Patty Kurtz, Jeff Kanipe, and my good friend Bob Naeye. Still more thanks go to the Italian translator, Giusi Galli, and to my ex-student and friend Letizia Stanghellini, who assured me that the translations were wonderful and who could call the Italian office to check on the articles' progress.

I would also like to thank the many providers of the images and figures who are credited in the legends to the illustrations. For brevity and simplicity, the names of organizations and instruments are commonly abbreviated, including the Associated Universities for Research in Astronomy (AURA), Associated Universities Incorporated (AUI), European Southern Observatory (ESO), Hubble Space Telescope (HST), International Ultraviolet Explorer (IUE), Jet Propulsion Laboratory (JPL), National Aeronautics and Space Administration (NASA), National Optical Astronomy Observatories (NOAO), National Radio Astronomy Observatories (NRAO), National Science Foundation (NSF), Space Telescope Science Institute (STScI), University of California Observatories (UCO), and Very Large Array (VLA). Every effort has been made to obtain permission to use copyrighted material, and sincere apologies are rendered for any errors or omissions. The publishers would welcome these being brought to their attention.

This book would not have been possible without the trust and patience of those at the Cambridge University Press, in particular Simon Mitton and Adam Black, who waited without complaint. Special thanks too to Brian Watts, who masterfully copy-edited the manuscript, to Sue Tuck, who efficiently controlled production, and to Zoe Naylor for her excellent design of both text and cover. Finally, thanks again to my wife Maxine who, as always, provided continuing encouragement and support.

James Kaler
Urbana–Champaign

Acknowledgements

# Prologue

At first the stars look alike, differentiated only by their apparent brightnesses. Some few are easily found from town; in the darkening countryside, you see more and more until seemingly endless numbers disappear into the blackened sky. Aside from the obvious differences in brightness, they are not all the same. Some display subtle shades of color from pale reddish through yellow-white to subtle blue, their hues enhanced with binoculars or telescopes. Here and there you can even find some that are quite noticeably red.

After sunrise, when our Sun clears the horizon and its haze, we see its yellowish light stream through a window and can make a connection with stars of the same color seen at night. Is the Sun similar to these? Can we see ourselves mirrored in the nighttime sky? And why do so many other stars not look quite like the Sun? These colors are our first hint that stars may differ from our own. What else might differentiate them, and more, to what degree might they differ?

Over the past 150 years, since the distance to the first star was measured, we have learned many of the answers. We have found stars that both dwarf the Sun and that are dwarfed *by* the Sun. Some are comparable to the orbits of the outer planets; others are the size of a small city. A few shine so brightly that if at the distance of the nearest star we could read by their light; others would have to be nearly in the Solar System to be seen at all. Scattered among these are stars so hot or cool that they glow mostly in invisible colors. And mixed with these are odd beasts so strange as to defy credibility.

Embark here on a journey of exploration to see the stars close-up, to witness a prodigious range of properties produced by differences in stellar masses and by stellar evolution, the inclusive phrase for a variety of aging processes, on which are superimposed bizarre effects produced by stellar duplicity. In their turn, we will look at limits of opposite ends of stellar behavior: at the brightest and faintest; the coolest and hottest; the biggest and smallest; and, most remarkably, see how Nature transforms one kind into the other.

# Chapter 1

# Sun and stars

Our ordinary Sun provides a baseline, a standard against which we compare other stars, against which stellar limits can be tested. Though it does not begin to approach the limits of stellar properties it is still a wonder. Even at a distance of 150 million kilometers (93 million miles) it provides sufficient light and heat for us to thrive. The Sun's properties amaze. One-and-a-half million kilometers (860,000 miles, 109 Earth diameters) across, it could hold a million planets like ours. At its gaseous "surface," its opaque "photosphere" from which sunlight streams to space, the temperature is nearly 6000 degrees kelvin. (Kelvin degrees, K, are Celsius degrees above absolute zero, $-273\,°C$. Throughout the text, "degrees" is commonly dropped and the temperatures expressed simply as "kelvin;" the Sun's temperature is therefore given as 6000 kelvin or 6000 K.) The fiery gaseous center reaches an awesome 15 million K at a density 14 times that of lead. The Earth's mass, measured from the strength of its surface gravity and radius, is $6 \times 10^{27}$ grams, 6000 million million million metric tons. The Sun, of which the Earth is a minor satellite, weighs in 333,000 times more, at $2 \times 10^{33}$ grams (the number derived from the Earth's orbital characteristics). The solar luminosity (the amount of energy our star produces as a result of compression through gravity and thermonuclear reactions in the heat of its core) is far beyond anything that humanity will ever produce. Shining with $4 \times 10^{26}$ watts, the equivalent of 4 million million million million hundred-watt light bulbs, it releases the world's annual energy production in one ten-millionth of a second. And it has been doing so for 4.6 billion (4600 million) years.

## Solar surface, solar light

Though unassuming when compared with other stars, the Sun has an outstanding characteristic that allows us think of it as extreme, as one at the edge: it is close to us, and we know far more about it than we do any other star, so much that theories cannot keep up with observational knowledge. To the eye alone, this magnificent

Figure 1.1. Stars seemingly pile upon stars in this photograph of the Milky Way in Cygnus. Seen from a great distance, our Sun would be just one of them, an ordinary star in the middle of an immense range of stellar properties. [From the *Atlas of the Milky Way*, F. E. Ross and M. R. Calvert, University of Chicago Press, 1934. Copyright Part 1 1934 by the University of Chicago. All rights reserved. Published June 1934.]

body appears as a perfect, featureless, yellow-white circle against the blue sky. (Do not, of course, try to look at the Sun or any solar feature without a professionally-made filter and a good knowledge of how to use it; exposure to full sunlight for even a fraction of a second can permanently damage the eye.) Use a telescope, however, and – as first discovered by Galileo in 1609 – a variety of features pop out. Toward the edge of the solar circle, the solar "limb," the Sun darkens noticeably. A closer look reveals the apparently smooth solar surface to be broken into thousands of tiny bright granules at the limit of vision. Make a movie and speed up the action, and the surface seethes with energy, the granules bubbling and boiling like a pot of oatmeal,

2

each tiny fleck lasting only a few minutes. The gases of the photosphere (and those far below) are in a state of convection, hot gases rising and losing their heat by radiation, cool gases falling.

The most obvious features are dark, seemingly black, spots set against the brilliant solar light. A few appear singly, but most of them are social, clumping into groups. Surrounding the spots are subtle white patches. The spots near the solar center are round, while those near the limb are distinctly elliptical. What appears as a disk is really a sphere, the spots near the edge appearing foreshortened. If you observe the Sun day after day, you find that the spots are not permanent features but come and go, new ones replacing old ones, some groups of them lasting a month, other simple spots a mere day. All, however, march steadily across the solar surface. This great body is rotating, taking 25 days at the equator for a full turn, but closer to 30 days near the poles, testimony to the Sun's gaseous nature, as a solid cannot behave this way. The Sun's average density (found from its mass and volume) is near that of water (one gram per cubic centimeter, 1 g/cm$^3$). A solid or liquid this massive would be much denser. The Sun must therefore be gaseous, not just at the surface, but throughout, even at its ultradense center.

The key to understanding the solar nature can be found on a summer afternoon. A thunderstorm flees to the east. Sunlight peeks from under the departing clouds and shines upon still-falling drops of rain, and a rainbow frames the sky, a circle of colors – red, orange, yellow, green, blue, violet – centered upon the point directly opposite the Sun. Yellowish sunlight actually consists of an array, a spectrum, of different colors that have been spread out by the light's passage through the raindrops. Isaac Newton created the same effect when he passed sunlight through a prism.

The explanation of the rainbow lies in the nature of light. Light behaves as a travelling electromagnetic wave with alternating electric and magnetic fields speeding along at 300,000 kilometers (186,300 miles) per second. It can also be thought of as a collection of particles – photons – that in a crude sense carry a chunk of wave along with them. Either way, light carries energy, and is the chief way energy is transported in

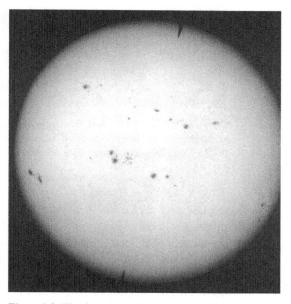

Figure 1.2. The Sun, 1.5 million km and 109 Earths across, contains 333,000 times the mass of Earth and 1000 times the mass of Jupiter. The spots, areas cooled by strong magnetic fields, are foreshortened near the solar limb, showing the Sun to be spherical. The darkening at the limb shows the solar gases to be slightly transparent and the solar temperature to climb inward. [Mt Wilson and Las Campanas Observatories.]

3

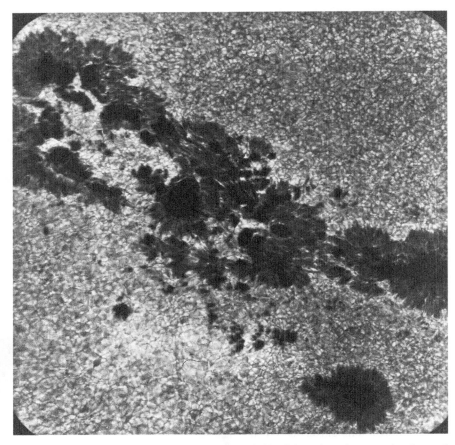

Figure 1.3. Sunspots have dark inner zones surrounded by lighter rings that are striated toward the surrounding photosphere, which is heavily granulated by convection. Convection is reduced in the spot by intense magnetism, cooling the gas and dampening the flow of radiation. [AURA/NOAO/NSF.]

the Universe. The color we associate with the light depends on the wavelength, the distance between successive wave crests. Red light, at one edge of the rainbow, has a wavelength of about $7 \times 10^{-5}$ cm (0.00007 cm), violet, at the other, a wavelength of $4 \times 10^{-5}$ cm. To rid ourselves of exponents, we use a more appropriate unit, the ångstrom (Å), $10^{-8}$ cm long. Red light therefore has a wavelength of 7000 Å, violet 4000 Å, the other colors falling in between.

There is no reason that Nature should stop radiating at these wavelengths; this range is just all we can see with the eye. Beyond red, infrared radiation is felt as heat; beyond even that, as wavelengths approach a millimeter, we call them radio waves and use them to broadcast information. Shorter than violet lies the ultraviolet, in the 100 ångstrom realm, X-rays, closer to 1 Å, the "gamma rays." The amount of energy carried by a photon depends on its wavelength, shorter-wave photons carrying more energy than longer-wave photons. Ultraviolet waves that get through the Earth's atmosphere produce burns and protective tanning; at shorter wavelengths radiation can kill; gamma rays, for example, are produced in atomic bomb explosions. Longer waves, however, are relatively benign: you can stand all day under a high-powered radio transmitter with perfect safety.

Any body with a temperature above absolute zero will attempt to radiate its energy away. Since temperature is a measure of the energy inherent in a body, the hotter the body, the greater its ability to radiate at more energetic wavelengths. At 3 K, all that is produced is radio waves; at 300 K, infrared (in addition to radio) is radiated, but there is insufficient energy to produce X-rays, which take closer to 300,000 K. Moreover, the greater the temperature, the greater the total amount of energy radiated. Around the turn of the twentieth century, this concept was quantitatively codified into a variety of radiation laws. A solid or pressurized gas (like that in the Sun) radiates a "continuous spectrum" that depends on its temperature. From a heated body all wavelengths down to a critical limit are present; a graphical representation shows no gaps, breaks, or jumps; as we ascend from longer to shorter wavelengths, the intensity (amount) of radiation first slowly increases to a peak at a characteristic wavelength then suddenly drops.

As temperature increases, two things happen. First, more radiation pours out at every wavelength, the amount being proportional to the fourth power of the temperature (double $T$ and the intensity of the radiation per unit area climbs by a factor of $2 \times 2 \times 2 \times 2 = 16$), a rule called the "Stefan–Boltzmann law." Second, and intimately related, the wavelength of maximum radiation shifts shortward in inverse proportion to temperature, the "Wien law." From either rule we can find $T$ from the spectrum, either by determining the total amount of energy radiated per unit area or by finding the position of the peak, the wavelength at which the object is brightest. We cannot of course make such a measure by just looking at the spectrum, but must use a device that can sense the actual amount radiated at each point, a "spectrograph."

These principles explain the colors of the stars in the nighttime sky, reddish stars cooler than white ones (which have their radiation peaking in the middle of the visual spectrum) and white ones cooler than bluish ones. They also explain the darkening at the solar limb. For the limb to be darkened, the gas of the photosphere must be somewhat transparent so that we look a short distance into it. Since the Sun is spherical, we do not see as deeply when we look away from the center as we do at the center itself, as our line of sight enters at an angle. Because the limb is darker (and redder as well), the higher-level gases must be cooler to radiate less energy. Limb darkening not only supports the concept that the Sun is gaseous, but also shows that temperature increases inward.

This temperature increase has a profound effect on the solar spectrum. When we look at the solar spectrum in sufficient detail, we find that it is *not* continuous. Crossing it are vast numbers of dark gaps that cut out extremely narrow bands, or "lines," of color. Over the past century, each of these lines has been identified with a specific chemical element or compound. In the simplest sense, any given chemical

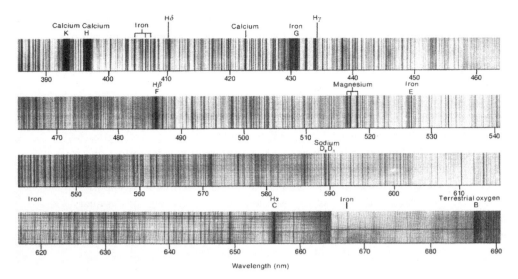

Figure 1.4. The solar spectrum, from the violet at upper left into the red at lower right, contains thousands of absorption lines of common metals. The scale is nanometers (1 nanometer = 10 ångstroms). The strongest lines are those of ionized calcium, followed here by hydrogen (H) and neutral magnesium and neutral sodium. The Roman letters B, C, F, H, K, L, etc. are from an older designation system. Three-quarters of the natural elements have been identified in the solar spectrum, and the others are surely there. [AURA/NOAO/NSF.]

element is made of an atom that has a central nucleus composed of protons, which carry positive electric charges, and neutral neutrons (particles with no electric charge, hence the name). Hydrogen, for example, always has a single proton, whereas helium has two, carbon six, iron 26, uranium 92. The nucleus is normally surrounded (loosely, orbited) by a number of negatively-charged electrons equal to the number of protons, rendering the atom electrically neutral. Since positive and negative charges attract each other, the electrons and protons are bound together. If any electrons are missing – a result of collisions between atoms – the atoms become positively charged "ions" that have absorptions completely different from those of their parent atoms and can therefore be uniquely identified.

The electrons of the atom or ion are responsible for the dark lines. As the radiation from a source of continuous light passes through a gas made of a particular chemical element, the electrons will absorb the photons, the electrons raising their own energies in the process. Because a given element has a particular electronic structure, absorptions relating to it will occur only at particular wavelengths associated with that element (or ion). Hydrogen produces only a few absorptions (in the red at 6563 Å, the blue at 4861 Å, respectively called Hα and Hβ, as well as a few others); helium, with two electrons, has many more, and iron has hundreds of thousands. But whatever the number, for a given atom or ion (or molecule, a combination of atoms that makes chemical compounds), they are always in the same place in the spectrum. We can therefore sense the presence of hydrogen or any element to the distant reaches of the Universe. In the Sun's photosphere, the deeper, denser, hotter gases produce a continuous spectrum that must pass through higher, cooler, less-dense gases that superimpose their absorptions. By comparing all the line positions with laboratory measurements, we find out what is in the Sun. Of the 90 or so natural elements that exist in the Earth's crust, we have found 68.

Some of the solar absorption lines are very strong, extracting great amounts of energy from sunlight; examples are singly-ionized calcium, $Ca^+$ (calcium atoms with one electron stripped away), neutral sodium, and hydrogen. Other rarer elements like cesium and tin have only very weak lines that extract little energy and are hardly noticeable against the colored background. The 20 or so elements seemingly missing from the solar gases must simply have lines that are too weak to see.

The strengths of an element's absorption lines depend only to some extent on the element's abundance. Of much more importance is the efficiency of absorption, which, for example, is much greater for ionized calcium than it is for hydrogen. The origins of the efficiencies lie in the atoms' electronic structures. Though the electrons are sometimes said to orbit the nucleus, they behave nothing like a planetary system. In the simplest sense, electrons can exist only in orbits that have specific energies and orbital radii that lie above minimum, or "ground," values. Think of the atom as a ladder. You can stand on the floor, or on any of the rungs (which are real, even though empty), but nowhere in between. It requires energy to climb the ladder. The farther you ascend, the more energy you expend and the more you can release when you jump down.

Electrons can climb the ladder when atoms collide or when they absorb photons from the flow of energy. The energy-rungs are responsible for the discrete nature of the absorption spectrum. The absorption of a photon of a specific energy – that of the energy difference between any two rungs – can make an electron go from one rung to another; when absorbed, that photon is removed from the flow of energy, and if enough are picked off by enough atoms, an absorption line is born. Rungs do not have to be next to each other for a jump to occur; the electrons can skip intermediate ones. As a result, a huge array of lines is possible. If the electrons jump *downward*, we see the opposite phenomenon: *emission* lines, *bright* lines of color at specific energies or wavelengths. Each kind of atom or ion has a different kind of ladder with a different number of rungs in different places. As a result, each kind of atom or ion has a different spectrum.

The optically-visible hydrogen lines can be produced in the solar photosphere only by electrons that are already on the second rung of the hydrogen ladder. Most people in the world stand on the floor; at any time only a few stand on the rungs of ladders. The same is true of atoms. There are so few electrons on the second rungs at any one time that the hydrogen absorption lines are weak even though there is a huge number of ladders. The ionized calcium lines, however, come from the floor, where the calcium ions have almost all their line-producing electrons. As a result, they can *all* absorb from the continuum. The efficiencies of absorption involve how

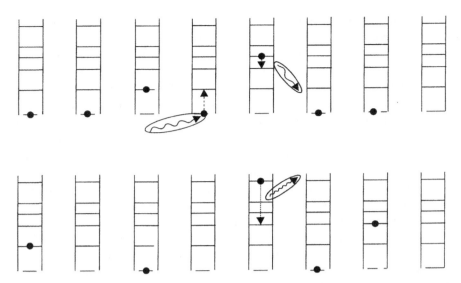

Figure 1.5. A collection of atoms is depicted as a set of ladders, the rungs of which increase upward in energy. Most electrons are on the floor, but a few (the number hugely exaggerated here) have been kicked up to various rungs by collisions. Electrons can jump downward with the emission of photons (wavy arrows) with energy differences equal to the energy differences between rungs; or they can absorb photons with the same specific energies. Photons trying to get through a complex set of atoms in the outer solar layers are absorbed at specific wavelengths to produce the solar spectrum.

many electrons at any given time are on whatever appropriate rungs multiplied by the probability that such an electron will actually absorb light.

When, after about 1920, astronomers were able to take these efficiencies into account, they found that the Sun consists primarily of hydrogen. Analysis of other data, including direct measurement of the matter that flows from the Sun past the Earth (the "solar wind"), has shown the Sun to be about 91% hydrogen and 9% helium. These numbers add to 100%; the rest of the elements are in the decimal places. Less than two-tenths of one percent is filled with all the other elements of nature. Oxygen leads, then carbon, neon, nitrogen, and the rest. There is no reason to think that most stars should be differently composed, and in fact for the most part they are not (though there are some wonderful exceptions).

## From the depths

The ever-present solar spectrum suggests a quiet peace, and the stars of the night-time sky almost define serenity. But the Sun, and by implication the stars, are anything but quiet, as shown by the changing granulation pattern and the ephemeral sunspots, which are wards of the solar depths, the convection of the outer third of the Sun creating them both.

Sunspots were watched for over two centuries before astronomers saw that they were cyclic: the number of spots on the Sun at any one time varies with an irregular period that averages 11 years. At the peak of the cycle we see hundreds of spots, at minimum they can disappear altogether. The spectrum helps us here too. If a gas absorbs radiation while it is in a magnetic field, the absorption lines will be split into twos, threes, or more (the "Zeeman effect"), the difference in wavelength between the components telling the strength of the magnetism. Sunspot spectra are split, magnificently so. The Sun has a global magnetic field somewhat like that of the Earth. But within the spots, the field strength increases to thousands of times terrestrial. Sunspots tend strongly to come in pairs that have different magnetic directions. In one hemisphere (as defined by the rotational equator and poles), all the pairs will be aligned in the same direction; in the other hemisphere, they are aligned oppositely. After the completion of the 11-year cycle, the spots switch directions, returning to their original orientations after 22 years have passed.

The electrons in a wire that is moving in a magnetic field flow with an electric current and a flow of electricity will produce a magnetic field. The solar magnetic field is similarly generated by movement of its ionized gases, by a combination of rotation and the deep convection. In turn, the "differential" rotation of the Sun (that it rotates faster at the equator than at the poles) seems to wrap up and concentrate solar magnetism. Convection locally lifts the field upward, forcing it to pop through the surface in great loops; the spots are formed at the points where the loops exit and enter the photosphere. The intense, concentrated magnetism inhibits the convection, chilling the surrounding area; as a result, the gases radiate less and appear dark against the photospheric background. The loops are highly unstable, causing the spots to change their structures; they can short-circuit each other and collapse, and thereby release their magnetic energy in vast explosive flares.

The magnetic energy generated deep within the Sun is responsible for creating a huge, enormously hot halo around the photosphere, the solar corona. At a temperature of two million kelvin, the corona's density is so low that it does not follow the Stefan–Boltzmann radiation rule, and is so dim that it cannot be seen against the blue sky. Only when the Moon covers the photosphere in a total solar eclipse does the pearly layer shine through. The corona is confined by the same kinds of loops that create the sunspots. Where the magnetism does not close it up, the thin hot gas easily escapes, in part responsible for producing the "solar wind" in which the Sun loses about $10^{-13}$ of its mass each year. The solar wind blows past the Earth at a speed measured in hundreds of kilometers per second. Coronal blobs released into the wind by collapsing magnetic fields can disrupt the Earth's field, generate intense electrical activity in the upper atmosphere, and create displays of the northern and southern lights. We are very much in the extended solar environment, our Earth beholden to what happens far below the surface.

Yet the true essence of the Sun lies even deeper, far below the convection layer. Limb darkening shows us that the temperature of the Sun (and by analogy that of any star) climbs as we proceed inward. Theory shows the same thing. Limb darkening gives us information on only the outer solar skin, while theory takes us deep inside, right to the center where we find the source of solar energy and support.

Figure 1.6. Magnetic loops that emerge from the Sun are made visible by confining (and heating) the thin, hot (two million kelvin) gas of the Sun's corona, from which blows the solar wind. [AURA/NOAO/NSF.]

A star is a battleground for the four "forces of Nature," those that act over a distance. The result of the contest is energy in the form of heat and light accompanied by the usually slow, but sometimes violent, aging of the star. The best known of the forces is gravity. Gravity, first described by Isaac Newton, draws all matter to all other matter, all atoms to all other atoms. It is also weakest of the four forces. We know it so well because it cannot be neutralized and because it acts over all space, its strength away from any mass decreasing according to the inverse square of the distance. It is the driving and organizing force of the Universe, acting to assemble matter, and is responsible for the creation of stars and their embracing galaxies.

The next one up in strength is the "weak force," which, unlike gravity, acts over only the size of the nucleus of the atom. It is responsible for various kinds of radioactive decay, in which one kind of particle, or one kind of atom, changes into another with the release of energy. Third is the electromagnetic force, which can manifest itself through electromagnetic radiation – light. Like gravity, it acts over all space but, unlike gravity, has two associated directions. The electric charge can be either positive or negative (the charges carried respectively by protons and electrons), and therefore electricity can neutralize itself. The normal atom contains equal positive and negative charges, and from a distance is neutral and safe. Only when the charges are unbalanced do we feel the power of the electromagnetic force directly (on an atom-to-atom basis $10^{35}$ times stronger than gravity), as anyone who has stuck a finger into a light socket will readily attest.

The greatest force of all is, by contrast to the weak force, the "strong force," which again acts over only the dimension of the nucleus. It is attractive in nature, and holds the particles of the nucleus together (and is thereby also called the "nuclear force"). Carried by both protons and neutrons, it is so strong that it can keep the nuclear protons (whose similar charges try to repel one another) clasped within its grip.

The balance of these forces makes the Sun work and give light to the day. A star contains enormous gravitational energy, its self-generated gravity trying perpetually to squeeze the gas together to make the star as small as possible. Gravity performs like the driving piston in an engine: as the gas is squeezed to higher density, it also heats. As we plunge into the heart of a star, any internal layer must carry an ever-greater load than the one above it, so it must be under higher pressure and hotter as well. Temperature therefore climbs as we proceed inward, the atoms moving ever-faster and becoming increasingly ionized as a result of violent atomic collisions. About three-fourths of the way into the Sun the temperature hits 10 million K. The speeds now become so great that even the repulsive force produced by their similar charges cannot keep them very far apart. A few can be driven so close that the strong force makes the protons stick.

Yet even the strong force lacks the strength to make two electrically-repelling protons join. During the brief moment the protons linger in company, one of them can release its positive charge via the weak force and become a neutron. The repulsive force suddenly disappears, and the two – the proton and new-born neutron – are bound by the strong force. The result is an "isotope" of hydrogen. The nucleus is still hydrogen because of the one positively-charged proton, but is a heavy version with an attached neutron. Since there are two particles now in the nucleus it is called hydrogen-2 (or $^2$H, where the number of particles in the nucleus is given by the superscript), and more commonly "deuterium."

The positive charge flies away from the nucleus as a positive electron, as anti-matter, normal matter with reversed charges. The Universe is mostly normal stuff. It has to be, as matter and antimatter cannot co-exist; they annihilate each other on contact with the release of energy. The positive electron – a "positron" – cannot get very far within the dense gas before it hits a normal negative electron and the two disappear. In their place appear two high-energy gamma rays. The Sun, through the compressive force of gravity and the actions of the strong and weak forces, has created energy from matter via Einstein's most famous equation $E$ (energy) $= m$ (mass) $\times c^2$ (speed of light)$^2$, the energy flying off, thanks to the electromagnetic force. Accompanying the positron is a near-massless (perhaps really massless) neutral particle, a "neutrino," that carries additional energy.

Almost as soon as the deuterium is made, another high-speed proton invades the nucleus and is captured by the strong force, which with three particles is now able to tie two protons together, creating an isotope of helium, $^3$He, as well as another gamma ray. Finally, two of these collide, resulting in $^4$He and the release of a pair of protons. In this "proton–proton" chain, four protons have melded into one atom of helium.

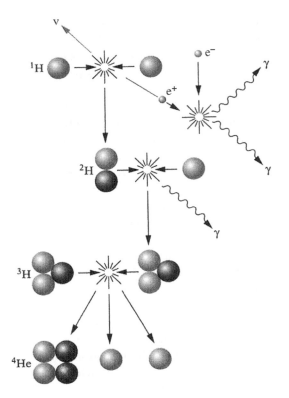

Figure 1.7. The center of the Sun is so dense and hot that four protons or atoms of hydrogen (the lightly shaded spheres) can fuse to become one atom of helium via the proton–proton chain. The chain first makes deuterium when a captured proton turns into a neutron (dark sphere) with the release of a positron (a positive electron), labelled $e^+$. The deuterium absorbs another proton to become light helium, $^3$He, which undergoes another reaction to make normal helium, $^4$He. Energy is released as neutrinos ($\nu$), near-massless particles that fly unimpeded from the Sun, and as gamma rays ($\gamma$) created by the collisions of positrons and electrons. [From *Cosmic Clouds* by J. B. Kaler © 1997 by Scientific American Library. Used with permission of W. H. Freeman and Company.]

Gamma rays are deadly, and life on Earth would be impossible if we were in their full glare. We are rescued by the Sun's vast outer envelope, the same one that raises the temperature of the core to the heights that make the fusion reactions possible in the first place. The gamma rays cannot penetrate the envelope directly. Instead, they are immediately absorbed by atoms and then re-emitted. Gradually the energy works its way through the outer layers. Since these are cooler than the inner layers, the emitted photons must on the average have lower energies. But since once energy is created it cannot be destroyed, there must be more photons to make up the difference. As a result, a single gamma ray created in the solar core will – after nearly a million years – result in the release of thousands of optical photons – those seen with the eye – from the solar surface. The neutrinos, on the other hand, speed silently and immediately from the solar center right to the Earth, where with great difficulty we can detect them for a direct "look" into the solar center and a confirmation that the reactions indeed take place as predicted.

## Other stars

The Sun is but one of 200 billion stars in our local collection, our Galaxy, if "local" is a term that can be used for a structure that is some $10^{18}$ km across. Such distances require the use of a larger unit. The light-year (l.y.), the distance a ray of light – a photon – travels in a year at a speed of 300,000 km/s, is $9.5 \times 10^{12}$ km long. The distance to the Sun, the "Astronomical Unit" (AU), is 150 million km (8 light-minutes), so the light-year has a length of 63,000 AU.

Decades of research have shown that our Galaxy is dominated by a thin disk 80,000 or so light-years across that contains over 90% of the stars. We are located

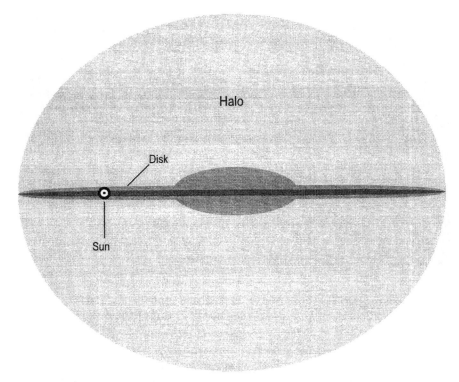

Figure 1.8. Seen edge-on, our Galaxy consists of a thin disk that contains most of the stars surrounded by a sparsely populated halo. The Sun is out toward the edge (about 25,000 light-years from the center). The disk encircles us as the Milky Way. The disk and halo faintly extend much farther than indicated here.

about 25,000 light-years from the center, rather well off toward the ill-defined edge. The disk rotates – the Sun taking about 250 million years to go about the center – and is structured into a set of flowing spiral arms. Surrounding the disk is a somewhat spherical, sparsely populated halo that the disk slices in half.

Ours is hardly the only galaxy, a suspicion confirmed by Edwin Hubble in the 1920s when he found that fuzzy blobs observed for centuries were distant vast collections of stars. They in fact swarm the Universe, tending strongly to clump into huge clusters. There are numerous kinds, but three broad varieties dominate. The loveliest are the spiral galaxies like our own. Since we cannot see ours from outside, other spirals tell us a great deal about the system in which we live. The elliptical galaxies on the other hand are seemingly simple ellipsoids that exhibit neither disks nor spiral arms. Another smaller fraction consists of irregular galaxies with little structure. Tucked in among them all are vast numbers of small assemblies that look like shredded debris. Given enough time the Hubble Space Telescope could probably detect a trillion (a thousand billion or a million million) galaxies, many much larger than our own.

Figure 1.9. Other spiral galaxies tell us how ours is constructed. On the left, NGC 891, presented edge-on, shows the thinness of its disk and its thick lane of interstellar dust. Seen more face on, the disk of M74 displays a complex set of spiral arms somewhat similar to ours. [*Left*: Mt Wilson Observatory. *Right*: AURA/ NOAO/NSF.]

Because the Sun lies within our Galaxy's disk, we see the disk and its billions of stars surrounding us in a great thick white band called the Milky Way. The subject of myriad mythologies, the Milky Way was revealed as made of stars by Galileo when he turned the first astronomical telescope on it in 1609. At its best, it is a spectacular sight that unfortunately is easily lost in the glare of artificial lighting. Its enormously complex structure is created by thick clouds of dark dust that lie in the spaces between the stars and that appear to divide the Milky Way into parallel tracks. This dust lane is easily seen in images of other galaxies set edge-on. The dust, allied with massive clouds of gas, blocks the light of stars. Within the clouds, the temperature plummets to near absolute zero, allowing the contraction of the gas into new stars. Stand out under the thickest parts of the Milky Way and look into its black hearts: stars are being created at that moment in the hidden darkness. Our own Sun came from such a cloud 4.6 billion years ago.

Aside from its observational accessibility (and its third, life-holding, planet), our Sun has no special characteristics that set it apart from the other stars of the Galaxy, and lies very much in the middle of the ranges of all stellar properties. Measurement of such properties for other stars has in one way or another occupied astronomers for 2000 years. The simplest of them is apparent brightness. About 150 B.C., the great Greek astronomer Hipparchus divided the stars into six brightness categories we now call magnitudes, the brightest (the top 21 stars) called first magnitude, the faintest the eye can discern, sixth magnitude. Nineteenth-century astronomers recognized magnitudes as a logarithmic brightness scale, and established a quantitative system in which first magnitude was exactly 100 times more luminous than sixth. If a difference of five magnitudes corresponds to a factor of 100 then one magnitude refers to the fifth root of 100, or 2.512 . . . (multiply it by itself 4 times). To calibrate the scale, the average of a collection of faint stars was arbitrarily set at 6.0 and the rest scaled to them.

Hipparchus's original first-magnitude stars contain a very wide range of brightness. Seven of the most brilliant (including Alpha Centauri, Vega, and Arcturus) had to be moved to magnitude zero, and two to −1. The scale is open-ended, allowing us to extend it to bright planets and to telescopically observed stars. With binoculars we can see to eighth, with a typical backyard telescope to perhaps 12th or 14th, and with the Hubble Space Telescope to 30th. Do not let the small numbers fool you; each set of five magnitudes is another factor of 100, so that 30th magnitude is a trillion ($10^{12}$) times fainter than Vega, itself shining at magnitude 0.03. Vega in fact now represents the modern standard, to which all stars are ultimately referred.

The apparent magnitude – the brightness the star appears to be in the sky – depends upon the intrinsic luminosity of the star and on its distance. The derivation of stellar distances begins with measurements of the parallaxes of the nearby stars, the minute shifts in position caused by the Earth moving in orbit about the Sun. If you look at any object from two points of view, it changes position relative to the background. Know the distance between the points and the angle of shift and you can calculate the distance. In the cases of stars, the angular shifts are so small that they were not measured until 1846; the largest such shift, for Alpha Centauri, is only 1.48 seconds of arc. (There are 3600 seconds of arc in a degree; for comparison, the full Moon is one-half degree across.) As a result, Alpha Centauri (actually a dim companion to it) is the closest star to Earth. The "parsec," the distance unit used in professional astronomy, is defined as the inverse of the parallax (formally, one-half the full shift) expressed in seconds of arc. The distance of Alpha Centauri is therefore $1/0.74 = 1.35$ pc. There are 3.26 light-years in the parsec, so Alpha Centauri is 4.4 light-years, or 280,000 AU, away; we see the star as it was over four years ago.

Though parallaxes were long restricted to the immediate vicinity of the Sun, modern technology, including the hugely successful Hipparcos spacecraft, extends the technique to over 1000 light-years, which defines a volume that encloses millions of stars of different kinds. We can then use these parallaxes to calibrate other distance methods, allowing us to work our way outward to distant limits of the observable Universe, in which distances are measured in billions of light-years.

Figure 1.10. The band of the Milky Way is the disk of our Galaxy seen from a vantage point out near its edge. [Akira Fujii.]

Figure 1.11. The southern hemisphere's Alpha Centauri (the left-most bright star), 11 degrees to the east of the Southern Cross (at right), lies only 4 light-years away. [Akira Fujii.]

Even the naked-eye stars cover an enormous range in distance, to far beyond a thousand light-years. Alpha Centauri, to us the third brightest star, is (as revealed by the telescope) actually a double star, two close stars in mutual 80-year orbit about each other on average separated by about the distance between the Sun and Saturn. The brighter of the pair is remarkably like our own Sun; by itself it would still be the third brightest star in the sky, but it is apparently bright (that is, bright to the eye) only because it is so close to us. The southern star Canopus is brighter in appearance despite being over 200 light-years away; obviously, Canopus is much the more luminous star. To know true stellar luminosities, we must have some way of removing the distance, and use a system of *absolute* magnitudes, $M$, which are the apparent magnitudes, $m$, that the stars would have were they at a standard distance of 10 parsec or 32.6 light-years. Since the apparent brightness of a pinpoint of light depends on the inverse of the square of its distance, we can easily calculate absolute magnitudes from apparent magnitudes once the distances are known.

Of course there are some complications in the magnitude scheme. The magnitude of a star also depends on its temperature and therefore also on its color. For example a star could be very luminous but so cool that it radiates mostly in the infrared, very little energy sneaking into the visible. The star would therefore appear quite red to the eye and (compared to other stars) relatively dim. Similarly, very hot, blue stars radiate a great deal of their light in the invisible ultraviolet. The apparent brightness of the star therefore depends on the color of the light in which we make the magnitude observations, which must be specified for the measurement to make any sense. Traditionally we use yellow light, which is appropriate to that seen by the human eye; such magnitudes are therefore referred to as "visual magnitudes," or $V$. The "absolute visual magnitudes," called $M_V$, are the astronomers' basic measure of visual luminosity. Total luminosity can be found and related to absolute visual magnitude through the star's temperature.

## Spectra: at the heart of it all

Star colors and temperatures are related to the stars' spectra. When astronomers first began observing stellar spectra (with simple prisms placed at the foci of their telescopes) in the early nineteenth century, they were confronted with a highly confusing situation: stars exhibit a wide variety of different kinds, of which the solar spectrum is but one example. Some, like that of Vega, were seen to be supremely simple, dominated by a progression of hydrogen lines. Others, like those of Capella, Aldebaran, and the Sun, were filled with metal lines, and still other spectra had the complex bands of molecules.

The initial step needed to understand the natures of the stars was the classification of these varied spectra. The first enduring scheme was developed in Rome by Father Angelo Secchi, who divided the stars into five types that are roughly comparable with the colors that can be distinguished with the human eye. Before the turn of the century, the observations, by then being obtained photographically, were good enough that a more refined system was needed. Developed at Harvard College Observatory in the United States by Edward C. Pickering, Williamina Fleming, Antonia Maury, and Annie Cannon, the scheme used Roman letters that originally ordered stars according to the strengths of the hydrogen lines, but which were actually based upon a variety of criteria. After several letters were dropped or merged into others, and the system reorganized according to the continuity of the appearance of all the absorptions, the spectral sequence familiar to all astronomy students – OBAFGKM – emerged. Recent discoveries have extended it to yet cooler stars and substars (those not massive enough to make the grade), these now included in new class "L."

While the A stars feature strong hydrogen lines, those in class B (where hydrogen is still strong) are possessed of neutral helium; in class O the ionized helium lines are strong. As we descend below class A, the stars develop strong ionized metal lines, then neutral metal lines, and in class M molecular bands dominate. It was obvious even to Pickering and Cannon that the system correlated with the colors of the stars and therefore with their temperatures, which we know run from about 50,000 K at the hot (O) end down to about 2000 K at the cool end of class M. The 6000 K Sun, rather in between, with strong ionized and neutral metal lines, is a G star. As spectroscopy improved, Cannon decimalized the letters to discriminate better between stars. The Sun is a G2 star, cooler than one at G0 but hotter than G5.

By 1930 the principles behind the sequence were understood. All the different kinds of spectra are produced by stars with about the same compositions, about 90% hydrogen, 10% helium, and commonly somewhat less than 0.2% everything else.

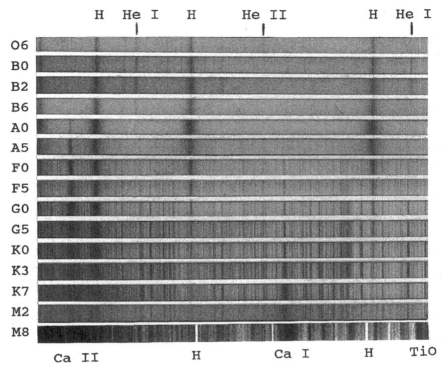

Figure 1.12. The standard stellar spectral sequence (*sans* L), seen here in the violet (between about 3900 Å and 4450 Å), descends from hot O6 (41,000 K) to cool M8 (2200 K). Hot stars have strong hydrogen lines that peak in class A then diminish downward, disappearing in class M. The hotter O and B stars display lines of helium, while the cooler F, G, and K stars show numerous lines of metals. The strengths of the singly-ionized calcium lines (one of which is blended with the left-hand hydrogen line) first increase with descending temperature, then decrease as the strength of the neutral calcium line increases. The bottom M8 spectrum displays a black band of the titanium oxide molecule. [*Atlas de Spectres Stellaires*, N. Ginestet, J. M. Carquillat, M. Jaschek, C. Jaschek, A. Pédoussaut, and J. Rochette, Observatoire Midi-Pyrénées and Observatoire de Strasbourg, 1992; bottom spectrum from *An Atlas of Representative Stellar Spectra*, Y. Yamashita, K. Nairai, and Y. Norimoto, University of Tokyo Press, 1978.]

(There are some fascinating variations on this theme that will be important later.) The differences do not just correlate with temperature but are *produced* by temperature, which changes both the efficiencies of line absorption and the ionization level of the stellar photospheric gases.

Look, for example, at hydrogen, whose lines are created in the outer layers of the star by absorption of photons by electrons on the second rung of the hydrogen energy ladder. As temperature rises, the gas has more internal energy as a result of the atoms' faster movements. There are therefore more electrons bounced upward to the higher rungs as a result of more vigorous atomic collisions. The hotter A stars have more electrons on the second rung of the hydrogen ladder than do solar-type G stars, and their hydrogen lines are stronger. Toward the bottom of the spectral

18

*The spectral classes*

| Type | Color | Temperature (K) | Description |
|------|-------|-----------------|-------------|
| O | bluish | 28,000–50,000 | ionized helium, hydrogen |
| B | bluish-white | 10,000–28,000 | hydrogen, neutral helium |
| A | white | 7500–10,000 | strongest hydrogen |
| F | white | 6000–7500 | hydrogen, ionized metals |
| G | yellow-white | 4900–6000 | ionized metals, hydrogen |
| K | yellow-orange | 3500–4900 | neutral metals |
| M | orange-red | 2000–3500 | neutral metals, molecular oxides |
| L | red | <2000 | neutral metals, molecular hydrides |

sequence, among the M stars, there are so few electrons on the second rung that the hydrogen lines disappear, even though the stars are still 90% hydrogen. Above about 10,000 K, the collisions are so vigorous that hydrogen's electrons can be ripped away to create hydrogen ions. The hydrogen lines (created only by the neutral atoms) therefore diminish in strength, though still remaining prominent right through the O stars.

Neutral helium's absorption lines arise from the second rung in its energy ladder as well. However, helium's second rung is twice as high as hydrogen's, and it takes much more energy – and higher temperatures – to kick an electron into it. As a result, we do not see helium absorptions in the Sun's photospheric spectrum; they in fact do not become visible until we reach the temperatures of the B stars. At yet higher temperatures, helium ionizes, its lines becoming prominent only in class O.

At high temperatures, metals are highly ionized, with two or more electrons missing. As we drop in temperature from the A stars, metals with only one electron missing become prominent, singly ionized calcium dominating the solar spectrum. Below class G, into K and M, the gas is no longer warm enough for the collisions to support even singly-ionized calcium, and its lines decrease in darkness, to be replaced by those of the neutral state (neutral calcium for example). Eventually, the temperature is low enough to allow molecules – combinations of atoms that are easily broken apart by collisions with atoms – to form. Even in the Sun we find a bit of CH and a few other hardy molecules, especially in the cooler sunspots. Among the cooler M stars (and in class L), molecules dominate, the cooler M stars recognizable by powerful bands of titanium oxide, TiO (which happens to have its absorptions in the optical part of the spectrum).

## Variety

The luminosities and temperatures of stars are traditionally presented on a graph in which absolute visual magnitude is arrayed against spectral class. The principal feature of this "Hertzsprung–Russell diagram" (or HR diagram, named after Ejnar

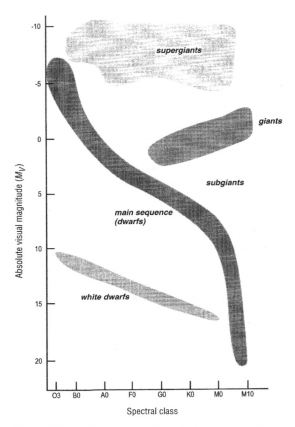

Figure 1.13. The HR diagram shows the luminosities of stars (expressed by absolute visual magnitudes) plotted against temperatures (expressed by spectral classes, excluding class L). The main (or dwarf) sequence runs from lower right to upper left. It is a mass sequence that starts at eight percent solar on the bottom to about 100 times solar at the top. The other zones represent various stages of stellar evolution in which stars are dying. Stars like the Sun become the giants and then the white dwarfs; higher-mass stars become supergiants and then explode.

Hertzsprung and Henry Norris Russell) is a strip densely packed with stars in which visual luminosity is in some direct proportion to temperature. That is, as the stars' surface temperatures increase, so do their absolute brightnesses. In observational terms, as we proceed through the spectral sequence, from M through G to O, absolute visual magnitudes decline from around +20 (a million times fainter than the Sun) to about −6 or −7 (over 50,000 times brighter). Such a correlation is in qualitative keeping with the Stefan–Boltzmann law, in which a hot body brightens according to the fourth power of the temperature.

To place these upper and lower luminosity limits in perspective, imagine one of these extreme bodies replacing the Sun. To light the day with a low-end star we would have to be 1000 times closer than we are to the Sun, or a mere 150,000 kilometers (about 100,000 miles), less than half the distance to the Moon. At the high end we would have to be over 200 times farther away than we are now, more than five times more distant than Pluto.

A star's luminosity – its power output – depends on two quantities. Temperature determines only the amount of energy radiated per unit area (square meter or square centimeter of surface). The more square meters of surface possessed by the star, the brighter it will be, so luminosity also depends on radius or diameter. The surface area of a sphere depends on radius squared. As a result, luminosity varies as $T^4$ times $R^2$. If we know the luminosity and temperature, we can find the radius. Luminosity is related to absolute visual magnitude; we must only be sure to factor in the invisible ultraviolet and infrared radiation respectively produced by hot and cool stars (which can be quite large). The stars in this "main sequence" brighten even faster with increasing temperature than mandated by the Stefan–Boltzmann law, showing that the stars are also increasing their surface areas, or diameters, from about only twice the size of Earth at the low end to over 10 solar diameters at the high end.

Hertzsprung's and Russell's greatest discovery, made in the early twentieth century, was that many stars do *not* lie on the main sequence. The principal additional feature of the diagram is a central band that goes up and to the right, in which luminosity increases as temperature *decreases*. To be both bright and cool requires great size. These stars were therefore rather naturally called "giants," discriminated by calling those of the main sequence "dwarfs." (In spite of the size of bright main sequence stars, the terms "main sequence" and "dwarf" are synonymous.) Giants can easily encompass the inner Solar System. We also see cool stars up at the top of the HR diagram that are even brighter than giants; these "supergiants" can encompass much of the *outer* Solar System. We also find stars *below* the main sequence, stars that are both hot and quite faint. These must be terribly small, even smaller than Earth itself. The first ones found were white, so the name "white dwarf" was applied, a term still used even though some are red and others blue.

These various stellar zones on the HR diagram were formalized in the 1940s by astronomers W. W. Morgan, P. C. Keenan, and E. Kellman, who placed them into luminosity classes distinguished by Roman numerals: I through V represent supergiants, bright giants, giants, subgiants (stars that fall between the giant and dwarfs), and main sequence dwarfs respectively. The Sun is, finally, a G2 V star.

Much of twentieth-century astronomy has involved the explanation of the HR diagram. The most important quantity is mass. In the long run, nothing much else matters. Masses are derived by the examination of double stars. The story starts almost 400 years ago when Johannes Kepler revealed the laws that govern planetary orbits. In his third law, he showed that the squares of the orbital periods in years were proportional to the cubes of their average distances from the Sun as expressed in astronomical units. (Jupiter is 5.2 AU from the Sun, and orbits in 11.9 years. Squaring the period and cubing the distance yield the same number.)

Newton derived the result theoretically from his laws of motion and the law of gravity and, moreover, found that the orbital period of a planet depends on both the distance of the planet from the Sun and on the sum of the masses of the Sun and the planet. If, for example, you could increase the mass of the Sun, but hold the Earth at 1 AU, the Earth would have to move faster as a result of the increased gravitational attraction, and the period would be less. Consequently, you can determine the sum of the masses of the Earth and Sun from the orbital characteristics of the Earth. Since the Sun is so much more massive than the Earth, the result is effectively the solar mass. Any of the other planets would serve equally well and give the same result.

We can apply the same reasoning to any two bodies in mutual orbit, and therefore to double, or "binary," stars. Ever since William Herschel confirmed the existence of double stars in the late 1700s, astronomers have learned that they are anything but unusual. A prime example is our closest star, Alpha Centauri. Perhaps 80% of the Galaxy's stars are in some kind of double system, the two orbiting each

Figure 1.14. Alpha Centauri is actually a double star, the pair locked together in gravitational embrace and orbiting each other every 80 years. The brighter of the two is remarkably like our own Sun. [AURA/NOAO/NSF.]

other. Multiples are common as well. We see systems in which a star orbits a double at a great distance (double Alpha Centauri has a distant faint companion), pairs of binaries that orbit each other (like Epsilon Lyrae), even (like Castor, Alpha Geminorum) binaries that orbit double-doubles.

Once we have measured the radius of a binary star's mutual orbit we can find the sum of the components' masses. One star does not actually orbit the other, however; instead, each member of a pair of stars swings about a common center of mass that lies between them whose location depends on the *ratio* of their masses. Once we have the sum and ratio, we can calculate the masses of the individual bodies. From hundreds of studies, astronomers found that the main sequence is a *mass* sequence, beginning at the bottom at eight percent the mass of the Sun (the minimum required to run the proton–proton chain), continuing upward through one solar mass in the G stars, to about 20 times that of the Sun among the B stars; theory extends the relation to over 100 solar masses among the O stars. The larger the mass, the more gravitational energy available, the greater the compression, and the higher the temperature at the core. As a result, high-mass main sequence stars are much more luminous than low-mass stars.

The second quantity needed to explain what we see in the HR diagram is stellar age. By analogy with the Sun, the entire main sequence is a stable zone of hydrogen fusion, nuclear "burning" (a common synonym for "fusion") supporting the star against the pressure supplied by gravity. Stars are remarkable self-regulating devices. As the fuel in the interior is consumed the core shrinks a little in response, which drives the temperature up somewhat and causes the remaining fuel to burn (fuse) somewhat faster and the core to eat slowly into the surrounding hydrogen. The result is that the star will reside for most of its life on the main sequence, only very slowly brightening and/or cooling at its surface as the fuel supply diminishes. The rate of change of position of the stars on the HR diagram is slight and serves only to give breadth to the main sequence and make it into a band. Stars near the left-hand edge are newly formed, while those near the right-hand edge have little time left to them.

When the fuel inside a star is finally used up, and core fusion shuts down, the star begins to die. The lifetime on the main sequence depends on how much fuel is available divided by how fast it is burned. Nuclear burning rates are so sensitive to temperature that high-mass stars live much shorter periods of time than do low-mass stars. The lives of lower-mass stars are so long that no K or M dwarf has ever had time to evolve off the main sequence in the whole 13 (or so) billion-year history of the Galaxy. An O star, on the other hand, can burn out in a few million years, one of the reasons that the high-mass O-type luminaries of a Galaxy are so very rare: they are ephemeral, and in a sense evaporate right before our eyes. Moreover, their births are intrinsically rare as well. For reasons not yet understood, Nature prefers to make long-lived low-mass stars, enough so that half the mass of the Galaxy is tied up in the dim red M dwarfs.

When the fuel is gone, gravity gets the upper hand and the core contracts, releasing gravitational energy, and perversely making the star temporarily brighter and the envelope larger. Stars like the Sun expand to become giants, those in the upper mass ranges becoming supergiants. The giants lose their extended envelopes and the cores become exposed as tiny white dwarfs. The supergiants explode in extraordinary blasts, "supernovae," that expel vast quantities of their matter into space leaving behind amazingly small bodies: neutron stars that are no larger than a small town, or even the fabled "black holes" that are so dense that nothing, not even light, can escape from them.

The final result is that even though two quantities – mass and age – are needed to describe the HR diagram, the second, age, also depends on mass, making mass supreme in the life of a star. The whole story of stellar evolution is one of a perpetual attempt of a star to contract, first starting with its condensation out of dusty interstellar gases. The main sequence is the first of many pauses and transitions along the way that give life and sparkle to the HR diagram. The stories of the main sequence and the other stages that follow will be told in the ensuing chapters by looking at extreme limits in a variety of categories, of stars at the edge, at the faintest, coolest, hottest, brightest, biggest, and smallest, as we see one extreme marvelously transform itself into another. With this setting in hand, we will then look at the outer limits of age, at the youngest and oldest stars, and finally at some of the stranger stars not already encountered.

# Chapter 2

# The faintest (and coolest) stars

The HR diagram is yet an abstraction that needs to be filled with the stars of the nighttime sky. The first step is to establish the reality of the main sequence. Before exploring the faintest stars nature has to offer, none of which is close to being visible with the naked eye, first look at what the easily visible stars can tell us. Most take us in the opposite direction, upward, toward the bright top of the main sequence, without which the faint bottom would make little sense. At the same time we begin to explore the temporal natures of stars and the Galaxy, into which we can then fit the stars all the way to Nature's edge.

## Upward

Begin with the Sun, at class G2 sitting squarely in the middle of the main sequence's band of dwarfs. By coincidence, the brighter member of the Sun's bright neighbor, Alpha Centauri, occupies nearly the same space. Now climb the main sequence stairway. The dwarfs of type F are not well known, though just before we move to the A stars we pass third magnitude Porrima, Gamma Virginis, which through a telescope appears as two F0 stars in orbit around each other.

Not until class A do we find the sky's familiar luminaries. Here visit first with A7 Altair in Aquila, the Eagle, the southernmost of the famed Summer Triangle, then with northern winter's Sirius (A1), in Canis Major (great Orion's larger hunting dog), at apparent magnitude −1 by far the brightest of all stars as seen from Earth and the southern point of the Winter Triangle. Leading the parade of A stars is Vega, class A0, in Lyra, the Lyre, which passes nearly overhead in summer in northern climes and defines the western apex of the Summer Triangle. With a mass three times the Sun's, Vega's absolute luminosity is over 50 times solar, somewhat greater than Sirius's luminosity. That Sirius is apparently brighter than Vega though actually farther down the main sequence vividly illustrates the role played by distance. Sirius is close to us, only nine light-years away, whereas Vega is 25 light-years

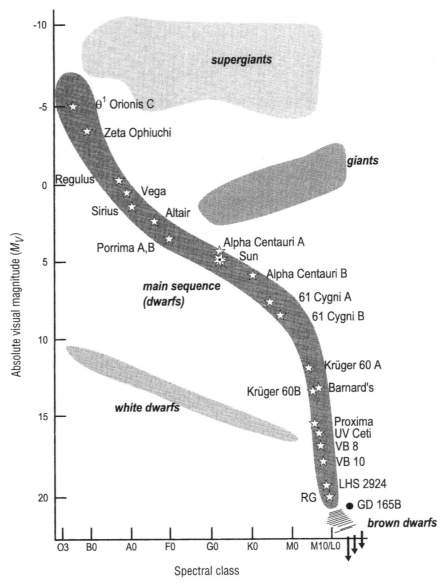

Figure 2.1. The upper main sequence is outlined by well-known bright stars. Fainter than the Sun, however, the stars become obscure, none of the dwarf M stars visible to the naked eye at all. Real stars seem to descend down to about absolute magnitude 20 or so and into a new class L, which contains the transition to the brown dwarfs, bodies too cool to be real stars. GD 165B is probably a brown dwarf.

distant. The A stars are also exemplified by the icon of the northern hemisphere, the Big Dipper (in Britain, the Plough). The middle five stars are all A dwarfs, in order from the bowl through the handle, A1, A0, A3, A0, and A2. The similarity of apparent magnitude tells us that the stars are all at nearly the same distance (about 80 light-years); in fact they are part of a small cluster through which the Sun is passing and that also contains Sirius.

On now to the B stars, defined well by class B7 Regulus, at the end of the "sickle" that represents the head of Leo the Lion, then to B3 Alkaid, the star at the end of the Dipper's handle. But now we begin to run out of bright stars. Second magnitude Zeta Orionis, the left-hand star of Orion's belt, at O9.5, just barely makes it into class O. To go farther, just to O9, we must drop to third magnitude and Zeta Ophiuchi (which helps define the southern boundary of Ophiuchus, the Serpent Bearer) and then into fourth magnitude to look at Xi Persei (O7). We need a telescope to see the end (or at least the near-end) of the main sequence, pinned down for us by O6 Theta-1 Orionis C, the brightest star of the "Trapezium," the group of stars that light the Orion Nebula, a giant cloud of interstellar gas. The main sequence actually extends at its hottest end to O3, near 50,000 degrees kelvin, but such stars are all so distant that they are faint and hard to see. Why, when the O brilliant stars top the main sequence, do we see so few? They are truly rare, and none happen to be close by, a topic to be revisited in the stories of the brightest and largest stars.

In tales of stars, their names become constant companions. Stars are named in a variety of fashions. Proper names derive from several different languages (mostly Arabic) and commonly describe the nature of the star or its place within a constellation. "Sirius," for example, comes from a Greek word meaning "scorching," certainly appropriate for the brightest star in the sky; "Altair" derives from Arabic and refers to "the eagle," for its place within the constellation of the Eagle. "Deneb," on the other hand, means "tail" in Arabic, as it (a class A supergiant) is at the tail of Cygnus the Swan.

Except for the brighter stars, proper names are cumbersome and hard to remember. About the year 1600, Johannes Bayer gave Greek letters to the stars within a constellation more or less in order of brightness, to which is appended the Latin possessive form of the constellation name. Hence, Altair, Sirius, and Vega are also Alpha Aquilae, Alpha Canis Majoris, and Alpha Lyrae.

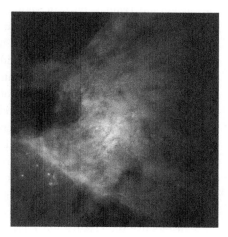

Figure 2.2. (See also Plate I.) The great Orion Nebula, viewed here by the Hubble Space Telescope, is lit by the four hot stars in the Trapezium at its center, the group dominated by Theta-1 Orionis C. [C. R. O'Dell (Rice University), STScI, and NASA.]

Two stars in Orion close together are named "Theta", and are distinguished by calling one "Theta-1" Orionis and the other "Theta-2." Theta-1 Orionis is the Trapezium quadruple star. The individual components of a double or multiple star system are labelled by Roman letters usually in order of visual brightness, those of the Trapezium called A, B, C, and D, hence Theta-1 Orionis-C that was discussed above as an example of an O star. (A, B, and D are also near the top of the main sequence, respectively classes O9, O9, and B1; the C component in this case is actually the brightest.)

There are only 24 Greek letters, so in the early 1700s, based on positional work by the English astronomer John Flamsteed, a good number of the naked-eye stars were numbered from west to east within a constellation, Vega becoming 3 Lyrae. Beyond these schemes, we simply use a variety of catalogue numbers that ignore the constellations and that usually take us around the sky from west to east from the vernal equinox, the mathematical point where we find the Sun on the first day of northern spring.

Without the Orion Nebula, we would hardly notice the Trapezium and the wonderfully luminous star Theta-1 Orionis. The nebula (Latin for "cloud") itself is a vast assembly of gas, mostly hydrogen, illuminated by ultraviolet radiation from the hot stars buried within (the nebulae will be explored in more depth under "the brightest stars"). The Orion Nebula is over 10 light-years across and contains a hundred solar masses of gas. A beautiful sight in even a small telescope, it marks a hotbed of star formation, the new hot stars within illuminating their birthplace. Such "diffuse nebulae" (there are several other kinds) abound within the Milky Way, and are major guideposts that lead us in learning how stars are born. Others of note are the Lagoon and Trifid Nebulae in Sagittarius, the North America Nebula in Cygnus, and the great nebula in the southern hemisphere's Carina. Orion's Trapezium provides an example of a different kind of multiple star system, its four bright stars making a close quadruple all roughly equidistant from each other. Unlike the double-doubles, this kind of system is unstable and will eventually break apart.

The Trapezium is but the pinnacle of a rich cluster of stars. The most obvious of such clusters is the Pleiades, over which Alcyone reigns. These clusters abound within the Milky Way, the disk of the Galaxy, thousands on view. The Pleiades, about 425 light-years distant, is compact and visible to the naked eye because of its bright bluish B stars. The Hyades, also in Taurus and seeming (by accident) to surround the K giant Aldebaran (Alpha Tauri), is closer, 150 light-years away, and thus angularly larger (almost to the point of making it seem not like a cluster at all). Closer yet is the one that contains the Dipper stars and surrounds us.

Clusters like the Hyades and Pleiades are loose and ragged. Known as "open clusters," they are typically 10 or 20 light-years across. Larger and vastly richer are the "globular clusters" that ignore the Milky Way and occupy the Galaxy's halo. They make up for their relative rarity – only some 150 are known – by their mass and luminosity. The greatest of them, Omega Centauri, contains over a million stars and is easily visible to the naked eye even though 17,000 light-years away.

The two kinds of cluster are differentiated by more than just appearance. A young cluster of stars should be born with an intact main sequence, from dim M through brilliant O. Since high-mass stars die first, the cluster's main sequence will burn away from top to bottom. (That clusters are commonly missing their upper main sequences is, in fact, powerful evidence that hot stars *do* die first.) We can, with the aid of the theory of stellar structure and the aging process (stellar evolution), date a cluster by where the main sequence stops. The Pleiades, with its B stars, is about 100 million years old; the Hyades, with no B stars, is closer to a billion. Open cluster ages range from near zero (those filled with O stars) to about 10 billion years. Since open clusters are part of the Galaxy's disk, we date the disk at about the same age. The globular clusters, however, *all* have stars missing above the cooler G subclasses, and therefore average about 13 or so billion years old. They are the oldest known systems, and since they are part of the Galaxy's halo, that too must be the halo's age; it had to have come before the disk.

We now have a more complete cosmic census. Next, look in the other direction, downward, along the main sequence, to the faintest of stars, which live so long that none has ever died.

## The lower main sequence

What is more awe-inspiring, the huge, perhaps unending Universe, or the almost-infinitesimal atoms that fill it with matter? Without the small, the large would have no meaning. We are tiny compared with the Galaxy, but have the power to comprehend its greatness. It is easy – but misleading – to confuse size with significance.

Figure 2.3. (*Left*) The Taurus's Hyades shine near the picture's center, the Pleiades near the right-hand edge. Both are open star clusters. The bright object is Mars, which happened to be passing through Taurus when the photo was taken. Open clusters are loose organizations with relatively few stars, as compared with the great globular clusters, epitomized by Omega Centauri (*above*) which contains over one million. [*Left*: author's photograph; *above*: AURA/NOAO/NSF.]

Look into the nighttime sky and see our brilliant stellar neighbors, perhaps Sirius or Vega, and you might quickly become accustomed to thinking they are somehow typical. Among the 40 brightest stars (counting close doubles as one star) 21 are giants or supergiants (or have them as parts of doubles). Of the remaining 19 main sequence dwarfs and subgiants (which are just above the main sequence but are not bright or large enough to be giants), eight are of spectral class B and eight others of class A, all substantially hotter, brighter, and more massive than our Sun. There is only one F star, first magnitude Procyon in Canis Minor (another piece of the Winter Triangle), and it is a subgiant. The remaining stars are the G and K combination that makes Alpha Centauri, stars that appear bright only because they are so close. Alone, Alpha Centauri's G2 star would shine at magnitude zero, and farther down the main sequence, the K1 component glows at visual magnitude 1.33. By itself it would still be the nineteenth-brightest star in the sky. But to find the apparently brightest star below Alpha Centauri B and the Sun, we must descend to fourth magnitude 40 Eridani. To the naked eye, the main sequence stops at the combined K5 and K7 components of 61 Cygni. And these are visible only because 61 Cygni, 11.4 light-years away, is one of the closest stars to the Earth (appropriately, it was the first to have its parallax measured). We see no main sequence dwarf M stars at all.

The lack of naked-eye lower main sequence stars is an example of the perfidious problem of "observational selection." We see first what Nature chooses to show us, including the bigger and brighter, while the smaller and dimmer are hidden from view. Astronomers spend vast amounts of time trying to overcome selection and account for its effects. Instead of looking just at the number of stars we can see, they must measure the number that lie within a particular volume of space (requiring knowledge of distance). The results yield a very different view from that of the eye alone. Though the number of visible bright A and B main sequence stars are the about same, there are actually five times as many A stars per unit volume as B stars. We see so many B stars because they are luminous and shine brightly over great distances, fooling us into thinking they are common. Though the O stars are even more luminous, they are so rare that all happen to be distant enough so that they practically avoid the naked–eye sky altogether: for every O star on the main sequence, there are 17,000 A stars!

Measurement, not just casual observation, shows that the number of stars increases quickly down the main sequence, in exact opposition to what we naïvely see. For every A star there are five F dwarfs, 13 G dwarfs, 15 of class K, and an amazing 100 M stars (1.7 million for every O star). There are more M (and L) dwarfs in the Galaxy than any other kind of star, all hidden by their dimness from naked-eye view. And even though their masses are small, typically a few-tenths solar, there are so many that they constitute about one-half our Galaxy's stellar mass. Through their combined gravity they therefore have a powerful controlling effect on the Galaxy's "dynamics," on the orbital motions of stars about the Galaxy's center, which give the Galaxy the appearance of rotating. The motions of the stars that we may have loved since we were children are heavily influenced by those of which we are largely unaware.

The red M dwarfs also can exhibit some distinctly odd behavior that is reminiscent of solar activity but on a scale that is relatively much grander, so that what we can learn of them we might be able to apply to our own life-giving Sun. Yet while the significance of the M dwarfs is now unquestioned, our knowledge of them has lagged behind that of their brighter cousins, a result of their feeble glow that can make them devilishly difficult to observe. So now in our trip downward along the main sequence, let us pass from the A and B stars (and giants) that make the patterns of our constellations into the shadowy world of the faintest stars to see what wonders they have to offer.

## Personalities

Rather than characterizing the M dwarfs as a group, look first at some individuals (see the table overleaf) arranged in order of descending luminosity. The number of doubles and multiples is impressive: only four stars are (seemingly) alone, and several are in triple systems. If you want to see any of these M dwarfs you will need at least a small telescope: then almost a dozen in the list are within your view. Notice how close all the stars are to the Earth: only three are beyond 30 light-years, typical of known red dwarfs in general.

Two red dwarfs have naked-eye companions that act as guideposts for easy telescopic viewing. At the top of the list, the M0 red dwarf Eta Cassiopeiae "B" is paired with a third magnitude G0 dwarf quite like the Sun (the "B" component is the fainter of the two in a binary system). The two are separated by 11 seconds of arc, and offer a striking color contrast. Here you can see both what our Sun might look like from a distance and at the same time get a vivid direct lesson in how the main sequence drops in brightness as spectral class advances toward the cooler end of the HR diagram. Eta Cassiopeiae B is five magnitudes – a factor of 100 – fainter than the brighter class G0 "A" component. Though only 19 light-years away, this most luminous of red dwarfs is well below naked-eye visibility.

The star 40 Eridani is more remarkable yet. The bright component is a yellow-orange fourth magnitude K0 dwarf. Over a minute of arc away is a companion that is itself a much closer double that consists of a 10th magnitude white dwarf (to be highlighted under the smallest stars) in mutual orbit around (and about 9 seconds of arc from) an M4 red dwarf some 1.5 magnitudes fainter. We now see better yet the drop down the main sequence: though 40 Eridani is 3 light-years closer to us than Eta Cassiopeiae, the M4 dwarf is still 3.5 magnitudes fainter than the M0 component of Eta. Third down on the list is the wonderful double Krüger 60 that consists of *two* red dwarfs (classes M3 and M4) in fairly close mutual orbit. The stars are near enough together (only about 9 AU apart) so that over the years we can easily watch them orbit one another, the cooler star the dimmer, as expected.

Two fainter dwarf M stars are justly famous, with features that far outshine their dim natures. Barnard's Star, named after its discoverer, Yerkes Observatory astronomer E. E. Barnard (1857–1923), is an M5 dwarf that glows redly from the

31

*Some red dwarfs*[a]

| Star | Apparent visual mag. (V) | Dist. (l.y.) | Spectral class | Temp. (K) | (M_V) | Mass[b] (Suns) | Comment |
|---|---|---|---|---|---|---|---|
| Eta Cassiopeiae B | 7.51 | 19 | M0 | 3800 | 8.7 | 0.19 | companion to white dwarf; flare star |
| 40 Eridani C | 11.17 | 16 | M4e | 3300 | 12.7 | 0.16 | flare star |
| Krüger 60A | 9.85 | 13 | M3 | 3500 | 11.9 | 0.27 | flare star, DO Cephei |
| Krüger 60B | 11.3 | ... | M4e | 3300 | 13.3 | 0.16 | highest proper motion |
| Barnard's Star | 9.54 | 6 | M5 | 3100 | 13.3 | ... | closest star; flare star |
| Proxima Centauri | 11.05 | 4 | M5e | 3100 | 15.5 | ... | flare star |
| L726-8A | 12.45 | 9 | M5e | 3100 | 15.3 | 0.10 | flare star, UV Ceti |
| L726-8B | 12.95 | ... | M6e | 2800 | 15.8 | 0.10 | flare star |
| Wolf 630A | 9.76 | 21 | M4e | 2600 | 10.8 | 0.4 | flare star |
| Wolf 630B | 9.8 | ... | M5e | 3100 | 10.8 | ... | VB 8 |
| Wolf 630C | 16.66 | ... | M7 | 2600 | 17.7 | ... | |
| BD+4°4048A | 9.12 | 19 | M4 | 3300 | 10.3 | ... | VB 10 |
| BD+4°4048 B | 17.38 | ... | M8 | 2200 | 18.7 | ... | |
| RG 0050 | 21.5 | 65 | M8 | 2200 | 20.0 | ... | |
| LHS 2924 | 19.7 | 36 | M9 | 2100 | 19.4 | ... | brown dwarf |
| Kelu-1 | 22.1 | 33 | L2 | 1900 | 22.1 | ... | brown dwarf |
| GD 165B | 24 | 103 | L4 | 1850 | 21.5 | ... | |
| Gl 229A | 8.14 | 19 | M1 | 3700 | 9.3 | ... | brown dwarf |
| Gl 229B | ... | ... | T | 1000 | ...[c] | ... | |

*Notes:*

[a] Arranged according to increasing spectral class or that of the faintest component

[b] Measured masses only

[c] Too faint to be measured

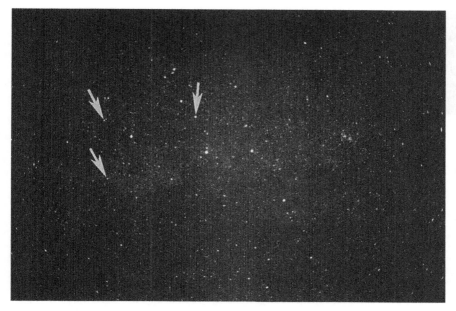

Figure 2.4. The familiar upside-down "W" of Cassiopeia is at center, the Double cluster in Perseus off to the right. The naked-eye part of the Eta Cassiopeiae (*right-hand arrow*) is a G dwarf. A small telescope will reveal an M2 red dwarf neighbor. Rho Cassiopeiae (*upper left arrow*) and 6 Cassiopeiae (*lower left arrow*), on the other hand, are among the Galaxy's brightest stars and will make a further appearance in Chapter 5. [Author's photograph.]

constellation Ophiuchus at magnitude 9.5. It beautifully reveals another property of stars. As they go about the Galaxy on their different orbits, they move relative to the Sun. Walk within a crowd of people hurrying down a sidewalk. Though going in the same direction, some walk faster or slower than others, and therefore you see people continually changing places. Occasionally, someone walks in the other direction, or charges through the crowd from a cross-street and whizzes by.

The motions manifest themselves in two ways. We observe people coming at or going away from us at some velocity – a "radial velocity" – or we observe them moving in the direction across our line of sight – a "proper motion." Astronomically, the radial velocities of stars are measured through the Doppler effect. When you move into a wave (or the wave into you), you see the waves coming more frequently, and the wavelength appears to shorten in direct proportion to your relative velocity. You can hear the effect on sound waves by listening to an airplane fly overhead: the pitch is higher as the craft approaches, lower as it recedes into the distance. In stars we see the effect in small shifts in the wavelengths of the absorption lines, and can therefore measure the stars' radial velocities, which are typically tens of kilometers per second, except for the rarer interlopers that move along at hundreds of kilometers per second.

Figure 2.5. In only 12 years (*left* to *right*), the M3 and M4 dwarf components of Krüger 60 make a 90° turn about each other. The difference that one spectral subclass makes in visual luminosity is striking. [Yerkes Observatory photographs.]

The velocities across the line of sight should generally be similar to those along it. The stars are so far away, however, that, even at such speeds, the resulting proper motions are insensible to the naked eye. To observe them we must usually watch and carefully measure positions for years, if not decades. But though the proper motions are typically measured in seconds of arc per century (a second of arc is a 3600th of a degree), the effect is cumulative. Come back in a million years and all the familiar constellation patterns will be gone. Proper motions depend on both distance and on velocity across the line of sight. If you stand by a highway, the faster cars move past your line of sight more quickly. But even a faster car on a more distant highway will appear to move slowly. If we combine proper motion, distance, and radial velocity, we can learn how the star is actually moving relative to us in three dimensions, and thereby can begin to construct a picture of how the Galaxy rotates and how the stars – including the Sun – move within it.

Barnard's Star holds the proper-motion speed record. Not only does it fly through space at nearly 140 km/s, but it is also only 5.9 light-years away, after the Alpha Centauri triple system the closest star to the Earth. As a result, it zips along at (for astronomy) an astounding 10.4 seconds of arc per year, a motion great enough to be visible with ease in a small telescope after only a few years. Barnard's Star draws closer to us (its radial velocity) at a speed of 108 km/s. From its proper motion and distance, its velocity perpendicular to our line of sight is 89 km/s, giving it an actual speed relative to the Sun of 140 km/s.

As Barnard's Star draws closer, its proper motion has to increase, an effect that has actually been observed: we can literally see the star moving through three-dimensional space. If the star had no proper motion it would be on a collision course with us. Instead, however, it will make only a close pass 10,000 years from now at a distance of 3.8 light-years, somewhat closer than Alpha Centauri is today. By then Barnard's Star will have moved northward out of Ophiuchus, through Hercules and Lyra, and into Draco (close to the north celestial pole) and will have brightened by a full magnitude (to 8.5, within range of large binoculars). At that time it will have increased its proper motion to 26 seconds of arc per year, moving a full degree, twice the angular diameter of the full Moon, in only a thousand years, afterward receding into the distance. The star vividly shows us that the heavens are not static, but alive with motion – we simply need time to see the effects. Barnard's Star was also at the center of a controversy that raged for 20 years over the possibility that it might have orbiting planets, a topic to be revisited below.

Though Alpha Centauri is usually called the closest star, it is actually a triple system that consists of the G2 and K1 double and a dim and distant M5 companion 2.2° away, which (appropriately called Proxima) is actually now the closest of all stars, 4.22 light-years away. With an absolute visual magnitude of but 15, it shines (if that be the right word) in our sky at only apparent magnitude 11, requiring us to use at least a 3-inch telescope (one with a 3-inch-diameter lens or mirror) to see it.

Though Proxima moves through space with Alpha and seems to be part of its family, there is still some debate as to whether Proxima is truly gravitationally bound to the bright pair. It may be on a "hyperbolic" orbit, in which it is falling around Alpha proper, but will someday leave at greater than the escape velocity never to return. If it is bound, and forever in orbit, it is now simply on the Earthward side and about 10,000 AU closer than Alpha itself. On the sky, however, Proxima is 2.2 degrees from its bright companion, so is actually some 14,000 AU away from Alpha. If the orbit it circular (which it probably is not), the period is over a million years.

As Barnard's Star approaches us, so do Alpha and Proxima. When Barnard's Star makes its close pass, Alpha will have closed to 3.6 light-years of us and it (rather Proxima) will *still* be the nearest star. It will continue to approach for yet another 17,000 years until it is only 3.1 light-years away, by which time the trio will have left Centaurus far behind (and will perhaps have acquired another name).

Proxima gives us a good chance to begin to appreciate the dimness of red dwarfs. The Sun's absolute visual magnitude is 4.83; Proxima is (in absolute terms) nearly eleven magnitudes fainter, a mere 1/19,000 as bright. From an Earth-like planet orbiting Alpha Centauri A at 1 AU, Alpha Centauri B (the class K dwarf) would make a brilliant orangy second sun. Proxima, however, would appear only as fifth magnitude, about as bright as the faintest star of the Little Dipper. It is not much to say for a companion to your star.

Yet this dim bulb of a star might have devastating effects. The Sun is surrounded by a vast cloud 100,000 AU or more in radius full of trillions of cometary bodies, small chunks of dirty ice that are the remnants of those that consolidated to make the outer planets

Figure 2.6. Only 6 light-years away, 11th magnitude Barnard's Star (*arrowed*) moves quickly (see text) against the background of far more distant stars. [© National Geographic–Palomar Observatory Sky Survey, reproduced by permission of the California Institute of Technology.]

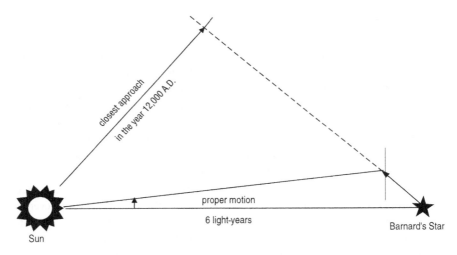

Figure 2.7. Barnard's Star is moving both toward us and across our line of sight with respective velocities of 108 and 88 km/s. The angular shift over the course of a year, the proper motion, is exaggerated here by a factor of 2000.

and that were kicked out of the planetary system shortly after the planets formed. Occasionally one works its way back in to appear in the sky as a great comet, the Sun melting its ice and blowing the resulting gas and dust into long tails. One may occasionally collide with the Earth, wreaking devastation. If Alpha Centauri produced similar bodies, imagine the effect wrought by Proxima as its gravity stirs the comets, sending vast numbers inward. Though it is problematic that Alpha itself has Earthlike planets because the action of the close pair may have prevented planet formation as we know it, other single stars with distant red dwarf companions would surely feel the effect, a steady bombardment that might make any kind of life impossible.

## Down to the bottom

Look now toward the bottom of the well. Not to be outdone by Krüger 60, Wolf 630 is a *triple* red dwarf that consists of an M4 – M5 double mutually orbiting an M7 star nearly 4 minutes of arc (1400 AU) away. This distant component has its own name, VB 8. With an absolute visual magnitude of 17.7, it defines the entryway to the end of the main sequence. Below it lie VB 10 (the "B" component to a double called BD +4°4048), LHS 2924, and RG 0050-2722, respectively M8, M9, and M8, with absolute magnitudes of 18.7, 19.4, and 20.0.

The names of these stars, unlike anything encountered before, give testimony to their faintness and to the number of astronomers who have spent years looking for (and cataloguing) them; moreover they give us a look into astronomical research activity that is rarely presented to the public. A. Krüger (1832–96) was a German astronomer who worked at the Bonn observatory helping Friedrich Argelander compile the enormous *Bonner Durchmusterung* (the "Bonn Survey"), the "BD" above, which catalogues a million stars counted in declination strips around the sky (declination the celestial analogue to latitude). Max Wolf (1863–1932) worked in Heidelberg, discovered numerous asteroids, helped develop the techniques of astronomical photography, and investigated dust clouds within the Milky Way. The "VB" in VB 8 and 10 refers to George van Biesbrock (1880–1974), long of the Yerkes

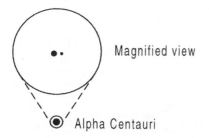

Magnified view

Alpha Centauri

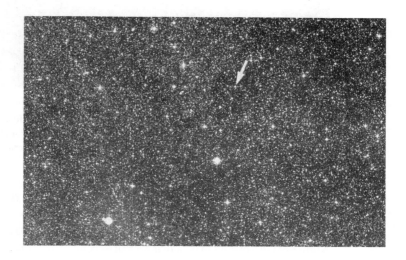

Proxima

To Sun

Figure 2.8. (*Inset*) The actual closest star to us, dim Proxima Centauri, is overwhelmed by brighter but much more distant stars. The sketch shows the spatial relationship. [ESO/SRC Southern Sky Survey.]

Figure 2.9. Comet Bennett, a brilliant comet that passed the Sun in 1970, was a brief visitor from the vast Oort comet cloud, possibly sent to us by the gravitational influence of some passing star. The bright tail was produced by both ionized gas and by dust released under the action of sunlight from the comet's small nucleus, which was only a few kilometers across. [University of Illinois Prairie Observatory.]

observatory, who was an expert in double stars and comets (and who in his 80s was known to run observing assistants ragged). "LHS" refers to "Luyten Half-Second," a catalogue of nearby stars that L. J. Luyten (1899–1995) of the University of Minnesota discovered to be moving a half-second of arc or more per year, and "RG" to the modern astronomers N. Reid and G. Gilmore. The number following their exceedingly dim star is a common designation that gives celestial coordinates (right ascension, the analogue to terrestrial longitude, and declination).

The M stars are cool enough to allow the formation of molecules, particularly titanium oxide. Molecular spectra are enormously complex. Unlike a single atom, a molecule can both rotate and vibrate, these actions splitting and broadening what would be a single atomic line into a whole series of bands, each one containing dozens, even hundreds, of individual absorption lines. A cool M star may have a million or more discrete absorptions. As the temperature drops from M0 to M8 or even cooler, more and more molecules form and the absorption bands deepen until the entire background continuous spectrum may be nearly eliminated.

Though the bands attain great strength in M giants and supergiants, the effect is more pronounced in the M dwarfs. Their small radii result in higher gas pressures

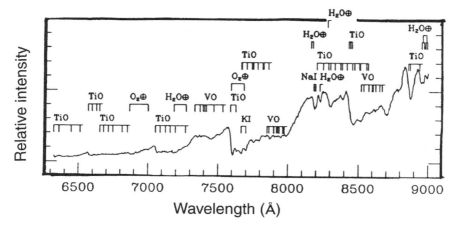

Figure 2.10. The spectra of red dwarfs are so complex it is difficult to analyze them theoretically. The absorption spectrum of the M9 dwarf LHS 2924, which includes titanium and vanadium oxides, blocks much of the emerging continuous spectrum. (The absorptions labeled with the crossed circle are produced by the Earth's atmosphere.) [J. D. Kirkpatrick, T. J. Henry, and J. Liebert, from an article in the *Astrophysical Journal*, 1993.]

that result in a greater proximity of atoms that forces more molecules into existence, rendering the spectra of the cool dwarfs more complex yet. Toward lower temperatures we see the signatures of three-atom molecules like calcium hydroxide (CaOH), and in the coolest M stars, of the vanadium oxide (VO) molecule. The general faintness of the stars complicates the matter further, making it quite hard to obtain high-quality detailed spectra over a range of types. Indeed, the whole concept of spectral classification becomes something of a problem.

The chopped-up nature of the spectra, coupled with the fact that at such chilly temperatures most of the stars' radiation is in the infrared (where it can be difficult to observe because of the Earth's atmosphere), makes temperature determination difficult. The Wien law cannot be applied accurately because the continuous spectrum is so terribly distorted by the lines, and we have to use atomic theory applied to the spectrum. The spectra of the M stars reveal cool temperatures that range from a high of 3900 K for Eta Cassiopeiae B downward to as low as 2100 K for LHS 2924.

Living near the bottom of the main sequence is a doubtful distinction; hardly anyone knows you are there, the result of very low mass. In order for RG 0050 (the faintest of the M dwarfs listed in our table) even to be visible to the naked eye, it would have to be a paltry 4000 astronomical units away, orbiting our Sun well within our comet cloud. If the Sun were replaced by RG 0050, it would shine not quite as brightly as the full Moon; you could barely read outdoors in full "dwarflight." These facts are a bit deceptive, however, since they involve only visible radiation. If we were to take *all* the radiation, here the infrared radiated by such a cool body, into account, Eta Cassiopeiae B would be about 1.5 magnitudes (about 4 times) brighter than we see it, and RG 0050 would be some 4 magnitudes, about 40 times, more luminous.

39

The temperature effect makes the faint dwarfs a little (but not *that* much) more respectable. It is this rapid displacement of the radiation to invisible wavelengths with advancing spectral type that causes the HR diagram's main sequence to seem to plummet so steeply downward in class M.

Were one of these dimmest stars to be our Sun, the effect would not help much with reading, but at least it would warm the day a bit more; not as if that much mattered, since we would receive only 1/32,000 the radiation we now get from the Sun, plunging the Earth into a deep freeze only a few degrees above absolute zero, the kind of heating experienced by comets in the cold Kuiper belt (the extended disk of the Solar System that contains one family of these primitive bodies) nearly 200 AU away from us, five times Pluto's distance.

The low luminosities of the dwarf M stars are a consequence of both low temperature, $T$, and small radius. The amount of energy released by a square centimeter of stellar surface is proportional to $T$ raised to the fourth power: at 3000 K it is only 1/16 that produced by the Sun. The total stellar luminosity is then equal to this value times the star's surface area; given the luminosity from the absolute visual magnitude (corrected for temperature), we can find the surface area and thus the radius. The stars are found to be quite small, and range from about 0.6 times that of the Sun at class M0 to only 4% solar – about half Jupiter's dimension and only four times the size of Earth – for RG 0050. (The greater stellar mass produces much greater gravitational compression and a body smaller than the much less massive planet; the star is nevertheless still gaseous.) At the Sun's distance, a star at the bottom of the main sequence would be only a minute of arc across, hardly discernible as a disk with the naked eye.

The most important characteristic of one of these stars (or of any star for that matter) is its mass, which controls the degree of compression, the internal temperature, and the luminosity. The larger the stellar mass the greater will be the heat in the core simply through gravitational compression. In turn, the temperature controls the rate of thermonuclear fusion, which keeps gravitational compression at bay. At the bottom of the main sequence, the mass hits the critical minimum that can take the central temperature above the seven-or-so million degrees kelvin mark required for full hydrogen fusion; below that, the hydrogen atoms (the protons) refuse to make deuterium, the first step of the proton–proton chain. We calculate from theoretical models of stars that the termination of full hydrogen fusion should take place at 0.08 solar masses, which represents the termination of the main sequence. This figure accords well with what we observe from binary stars.

Their light may be feeble, but the stars make up for it by being incredibly durable. The fuel consumption rate and the luminosity drop *very* rapidly with mass and hence core temperature. The Sun originally had enough hydrogen to last for 10 billion years, the ages of the meteorites showing that it has gone through about half that interval. At 0.8 solar masses (about spectral class G8), the projected lifetime is up to about 13 billion years, the age of the Galaxy itself. Below this mass, no star has ever had time to die, even if it is a charter member of the galactic club. Down at the bottom of the main sequence a star will live for trillions of years, trading the fame and glory of great luminosity for extraordinary stability.

What would life be like on a planet belonging to a star such as LHS 2924 or RG 0050? To be warmed by the same amount of heat, our planet would have to have an orbital radius of only a million kilometers, just 30% larger than the radius of the present Sun. Applying Newton's version of Kepler's laws of planetary motion, our "year" would be only 13 Earth-hours long. Tides would long ago have slowed our rotation such that one side of our planet would face the little red sun forever, much as the Moon faces us. It seems unlikely that life could exist.

## Active stars

So far, stellar (if not intelligent) life in the lower depths of the main sequence looks quiet and predictable, perhaps even boring. A large number of these stars, however, harbor a secret revealed only upon detailed examination. Take your telescope out-doors to watch Krüger 60. The fainter one, at class M4, is normally 1.4 magnitudes dimmer than its M3 companion. But if you are lucky, you might see Krüger 60B suddenly brighten. For several minutes it doubles its normal luminosity, becoming nearly as bright as Krüger 60A. As a consequence, Krüger 60B has taken on a new name, DO Cephei, indicating it to be *variable*.

The apparently serene night sky belies great stellar activity, even violence. The first actual "variable stars" found, stars that change in brightness, were of the explosive variety. Tycho's Star (now known to be a white dwarf star that blew up), reached the apparent brightness of Venus in 1572, then faded away, never to be seen again. Then in 1596, David Fabricius found Mira, a giant star located in Cetus that regularly disappears then reappears over a period of about a year. Huge numbers of regularly – and irregularly – varying stars are known. They fall into seemingly endless categories, the most prominent of which are those of the Mira class and the "Cepheids," named

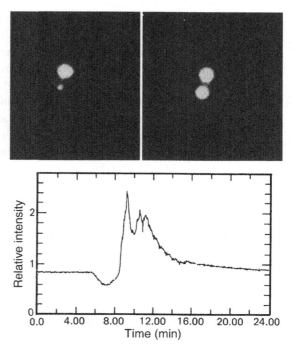

Figure 2.11. (*Upper, right*) DO Cephei (Krüger 60B) flares up, becoming as bright as its more massive companion. (*Lower*) The light curve (a graph of brightness against time) of the ultraviolet light from the flare star EQ Pegasi shows a precursor dip, a sudden brightening, and considerable subflaring. [*Upper*: Sproul Observatory photographs; *lower*: from an article by M. S. Giampapa *et al.* that appeared in the *Astrophysical Journal*, 1982.]

after the naked-eye star Delta Cephei, which varies between apparent visual magnitude 3.6 and 4.3 over an interval of 5.4 days, over and over again. These stars are pulsators: internally unstable, they change their radii and surface temperatures, the Cepheids with great regularity, the Miras with modest regularity. Other kinds of stars vary erratically, even unpredictably. Variables are important players among the cast of the extreme stars, and many kinds will be encountered as we proceed.

Only the brighter variables carry common proper or Greek-letter names. When Johannes Bayer ran out of Greek letters in his star-naming scheme, he continued with lower-case Roman letters and then with upper-case Roman letters. These are rarely used today, echoing only faintly through names such as "h and Chi Persei," the alternative name for the Double Cluster in Perseus, twin (naked-eye) open clusters loosely bound to each other. Bayer went through letter Q and stopped. When new variables continued to be found that did not have common names, the observers started with R and went to Z (in conjunction with the constellation name), then from RR to RZ, SS to SZ, up to the top of the alphabet to AA, etc., finally – after 334 letter combinations – to V335 and up. At least you can almost always tell a variable by its name.

DO Cephei was the 201st variable found in Cepheus (not counting four with Greek-letter names). Because of its penchant to brighten so suddenly, it is called a "flare star," or a "UV Ceti star" after the prototype discovered in 1947, although the first such flare was actually recorded (for DH Carinae) in 1924 by Ejnar Hertzsprung himself. Flare stars are remarkably common. Of the red dwarfs listed in our table (of which only UV Ceti itself was chosen *because* it flares), half produce these events; even Proxima Centauri belongs to the crowd. With a few K dwarf exceptions, they are confined to class M, and become more common down along the main sequence, at least to M5. The brighter ones are accessible to an amateur with a fairly large telescope, and with patience could provide an entertaining evening.

In a typical flare, the star brightens within in a few seconds, increasing from a few tenths of a magnitude to well over a magnitude, then takes some minutes to settle back to near, but not quite, normal; complete recovery may take several hours. Some flares, like that illustrated for EQ Pegasi (class M6, 21 light-years distant), show considerable sub-flaring, or even a dip before the flare. The individual events are erratic and unpredictable. A star may exhibit a tiny spark of only a few hundredths of a magnitude and next pop an enormous firecracker of a magnitude or more. The frequency of flares can be quite high, but depend upon the brightness considered. Little microflares can occur a few times per hour, but we might wait for 100 hours to catch a big one – and then bag two in a row. Consistent with their growing population toward cooler classes, the number of flares per hour is also greater down the main sequence.

These detonations are easier to see at shorter wavelengths: UV Ceti may brighten by one magnitude visually, but by two in the near-ultraviolet, indicating the flares' energetic natures, in which the temperature can climb to a million degrees kelvin or more (the temperature in the flare, not of the star itself). Wolf 424 is a champion, exploding by four magnitudes – a factor of 40 – in this wavelength band. These enormously energetic impulses are not confined to the narrow part of the spectrum seen from the ground with optical telescopes. We observe bursts in the radio spectrum, in the energetic ultraviolet, and even in the very high-energy X-ray spectrum (the latter two spectral regions by means of orbiting satellites).

There is yet another class of flare stars found especially in young clusters and in looser (gravitationally unbound) groups known as "associations." Their flares are quite energetic, run to earlier classes in type K, and have much longer rise and decay times: half-an-hour to peak (as opposed to seconds or a few minutes) and up to 10 hours to return. "Field stars," those not allied with clusters, have a mixture of ages (having come from all parts of the Galaxy), and tend to be considerably older than the cluster stars, revealing a link between flaring and age.

Almost all flare stars exhibit emission lines in their spectra and have the letter "e" appended to their spectral types. The stellar emissions indicate the existence of a deep chromosphere, a cool transition layer that lies between a star's photosphere and its corona. This thin layer is seen as a narrow red band of light surrounding the totally eclipsed Sun. The link between the Sun and the flare stars goes yet deeper. If we watch the Sun's chromosphere for a time, especially during a maximum in the 11-year solar activity cycle, we will observe an occasional large *solar* flare, the result of the magnetic loops that create sunspots collapsing or shorting out. These produce vast quantities of radiation, all the way from radio wavelengths to X-rays. Over a period of its hour-or-so lifetime, a typical big solar flare will generate almost a tenth of the energy that the whole Sun produces per second.

We see an apparently similar process in the red dwarfs, except that it is much, perhaps 100 times, more energetic. Instead of being the local phenomena we see on the Sun, the magnetic fields (which are comparable to sunspot fields in strength) and flares involve the whole star. Clearly life on an orbiting planet would be terrifying, if not impossible. Try to imagine sun-bathing (or dwarf-M-bathing) on the beach and having your star suddenly, with no warning at all, become 10 times brighter. Satellite observatories show us a steady stream of X-rays coming from some M dwarfs. Since X-rays are a product of the solar corona, we have direct evidence that the dim stars possess this "crown" as well. Moreover, there is a subset of flare stars (that actually extends into class K) called BY Draconis stars that vary by several tenths of a magnitude in a manner consistent with large starspots that rotate in and out of the field of view. We see then, down near the bottom of the main sequence, a repeat performance of solar activity, but now on a grand and wonderful scale; given the low masses of these stars, it is just the opposite of what we might naïvely expect.

The origin of the phenomena appear to lie in stellar structure and rotation. Stars from class F down along the main sequence have central regions that are in "radiative equilibrium." That is, the gas that makes up a star's core is quiet and does not circulate; energy is transported by radiation. But in the outer envelope, this equilibrium breaks down, and the convection that is responsible for the granulation of the solar photosphere sets in. All these stars rotate as well, the spin speed declining along

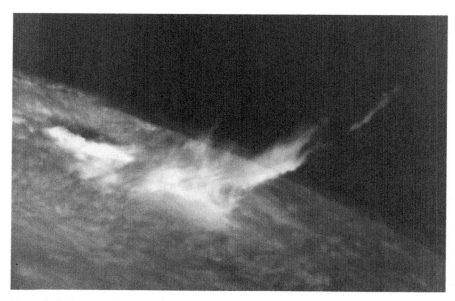

Figure 2.12. A brilliant flare erupts on the Sun. [AURA/NOAO/NSF.]

the main sequence from a few tens of kilometers per second at mid-F to only one, two, or a few kilometers per second in class M. (We can derive rotation speeds from the action of the Doppler effect on spectrum lines or from variation periods as derived by observing the effects of starspots going in and out of view.)

In the standard picture, the act of rotation on the convective envelope causes the circulation of ionized gas and acts as a giant dynamo that creates the magnetic field: the same field that loops out of the star to produce the flares. The fields are most likely anchored below the point at which the convection begins, for the Sun at least one-third the way in, within a region that is stable and transports the luminosity of the core by radiation. As the field extends outward from the stellar surface it causes a drag and acts as a brake to slow the star down. Consequently, as stars age, they tend to spin slower (though exceptions abound). The energies of the flares thus respond to the degree to which the rotation has decayed: the slower the star rotates, the less energy it has to give up, the less becomes its activity, and the corona, emission lines, and flares finally disappear. Flare stars are therefore the younger, relatively newly-formed M dwarfs, not those born at the time of galactic formation, explaining why flare stars are so common in young clusters. We can date a star, even if crudely, by its behavior.

But all is not well. Flare energies are higher than expected. Worse, theoretical models show that in dwarfs with masses under 0.3 times that of the Sun, corresponding to spectral class M4, convection takes over the star completely; even the core becomes convective, removing the anchor for the magnetic fields. Theory is supported by observation in that main sequence stars below this point are found to be rotating faster than those above it, showing that the magnetic braking effect is weakening. Yet these dim dwarfs still exhibit chromospheric activity and flares, and thus must maintain some kind or magnetic dynamo. We do not yet know why or how.

## Below the bottom

Beneath 0.08 solar masses there can be no stars as we ordinarily define them: stable self-luminous spheres of gas that are supported by full thermonuclear fusion. What happens below this critical value, in the mysterious depths under the bottom of the main sequence? Here we might see failed stars that should glow for a time after their formation as a result of heat generated by gravitational contraction and from a small bit of nuclear fusion. These "brown dwarfs" are not hot enough inside to start the proton–proton chain, in which two protons hammer together to create deuterium (hydrogen with an added neutron). But they are hot enough to start in the middle of the chain with their small amount of natural deuterium and for a brief time fuse that into helium. The brown dwarfs then cool, shrink to the size of planets (though with much more mass), and disappear from view. Though faint and difficult to capture, brown dwarfs are important for a variety of reasons.

Examination of the gravitational effects of our Galaxy and of the countless others scattered throughout the visible Universe show that there is much more mass than can be accounted for by stars and interstellar gas and dust. We are thus faced with finding it, this so-called "dark matter," mass that does not seem to radiate. It may constitute 90 or more percent the mass of the Universe. All sorts of suggestions have been proposed, from weird atomic particles through unobserved black holes to neutrinos (the near-massless particles created in atomic reactions) released during the Big Bang, the creation event of the Universe. Another possibility might be huge numbers of dim red or brown dwarfs or other small bodies. Dwarf M stars dominate the stellar universe. If their numbers keep rising from M0 through M9 and then keep climbing below the end of the main sequence, we might account for much of the dark matter. Moreover, the number of brown dwarfs would give us more information on the way stars are formed from dark interstellar clouds. They also – as we will see later – provide something of a bridge to less massive planets, with which they might overlap in mass.

The masses of bottom-end stars and brown dwarfs are so small that to make any impact on the dark matter problem, there would have to be huge numbers of them. However, there is some evidence from star counts, which are very difficult to make accurately, that the number of red dwarfs peaks at around class M5 and toward cooler classes begins to drop. Deep examination of our Galaxy's halo by the Hubble Space Telescope shows the same thing, a remarkable absence of faint red dwarfs. If the trend continues downward, we might expect relatively few brown dwarfs, making these exceedingly dim stars yet harder to find. Worse, it is difficult to discriminate between them and real red dwarfs at the bottom of the main sequence.

The discovery of brown dwarfs long presented a challenge. The list of discarded candidates that turned out to be faint stars, or not really even there, is quite large.

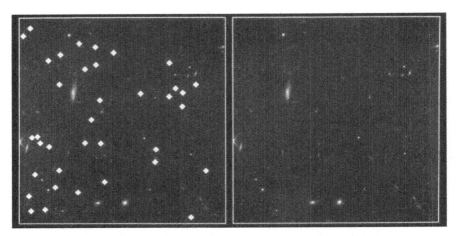

Figure 2.13. (*Left*) Diamonds superimposed on a Hubble Space Telescope image taken deep in to the Galaxy's halo show the expected number (not the location) of red dwarfs. They are absent from the un-retouched picture (*right*) suggesting that dark matter is not made of red, or even brown, dwarfs. [J. Bahcall, Institute for Advanced Study, Princeton University, STScI, and NASA.]

The most obvious way a brown dwarf might reveal itself is by its mass. If we can find a faint body in orbit around a real star, we can learn its mass by the application of Kepler's laws. However, brown dwarfs are so faint that just to see them would generally require that they be fairly distant from their companion stars, rendering the orbital periods rather long.

Other probes are more successful. As mass and luminosity drop, so does the surface temperature, making low-end stars (and brown dwarfs) so cool that they radiate most of their light in the infrared. Cameras designed to observe in the infrared reveal them easily, allowing the discovery of bodies so cool that astronomers have had to invent a whole new spectral class for them, class L (a letter originally used by Pickering and co-workers, but dropped as un-needed). In keeping with classical spectral taxonomy, the "L stars" are characterized not by temperature but by spectral characteristics. They are so cool that the titanium and vanadium oxide bands weaken and disappear (as the metals condense onto solid grains) and are replaced by metallic hydrides, chromium hydride (CrH) and iron hydride (FeH) becoming prominent. Absorptions of neutral metals – sodium, rubidium, cesium, even lithium – become important as well, depending on temperature, and the L stars decimalized like the others. Temperatures of stars with such complex spectra are difficult to measure. The coolest M stars hover around 2000 K. At L2 the temperature seems to be around 1900 K, dropping to perhaps as low as 1800 K by L4 and to 1500 K at L9. Are these brown dwarfs?

Lithium is easily destroyed by nuclear reactions at relatively low temperatures. All stars are endowed with an initial amount delivered from the interstellar gases from which they were born. But when it is swept into stellar depths by convection, the lithium disappears. The presence of the strong absorption lines of neutral lithium tells us either that the star is young (there is very little left in the Sun) or that the body is too cool to have destroyed it, suggesting that the star may be a true brown dwarf. The problem is that the candidate body might also just be young, so the star either has to be in a cluster of known age, or must have a companion whose age can be dated from the *lack* of lithium. A few stars in the Pleiades cluster qualify in that they have more lithium than we would expect on the basis of the cluster's lithium content, as do some others. About half the L stars, the cooler ones, show lithium absorptions and are probably brown dwarfs. The warmer L stars may be a mixture of brown dwarfs and real stars, the end of the true main sequence lying just below 2000 K in warmer class L. The L dwarfs therefore seem not only to be the dimmest stars but the coolest as well, just a bit chillier than their rivals, the extreme M giants.

Far below these lies Gliese 229B, an extraordinarily dim star whose spectrum is so different from the others that it was accepted as the first truly confirmed brown dwarf. The spectrum, which shows water vapor, and remarkably – like the Jovian planets – methane, suggests a temperature of no more than about 1000 K. Such bodies lie far below even the L stars, and class "T" has been proposed for them, as more are steadily being found.

Even if the two stars of a pair are so close together that they cannot individually be seen, we can still discover duplicity – and brown dwarfs. A close-double's spectrum will be a composite of the spectra of the components. As the individual stars orbit, they move alternately toward and then away from us. The Doppler effect shifts the spectrum lines back and forth, allowing us to measure orbital speeds, or, since we do not know the tilt of the orbit, lower limits to the speeds (if the orbit's plane is perpendicular to the line of sight, there are no back and forth motions at all). From such data, we derive lower limits to the masses of the components of binary stars. Imagine that one star is too faint for its spectrum to be recorded. It will still show up as it shifts the lines of its brighter companion back and forth. Such observations, coupled with the likely mass of the observed component (found from the known relation between mass and luminosity), allow us to infer a lower limit to the mass of the invisible star. If this limit is substantially below 0.08 solar masses, we can stake some claim to having found a brown dwarf. Many stars have now been so found, with masses of at least 0.013 solar, one-sixth that of the limit at the bottom of the main sequence.

Figure 2.14. (See also Plate II.) The brown dwarf Gliese 229B glimmers red next to its vastly brighter M1 dwarf companion. At a temperature of only about 1000 degrees kelvin, its infrared spectrum contains absorption bands of methane, which are never found in stars. [S. Kulkarni (Caltech), S. Durrance (Johns Hopkins University), STScI, and NASA.]

Though a few of these bodies might be stars or even heavy planets, there is no longer much doubt that brown dwarfs really exist. Just how common they are, however, remains to be seen.

## Planets

The mass of 0.08 solar masses that defines the bottom of the main sequence is 80 times that of Jupiter, and as we descend the mass ladder of the brown dwarfs, we approach the realm of the bodies in our own Solar System we know as planets. At what point does a brown dwarf cease being such and become a planet of its parent star? Are planets really just little failed stars, the dimmest of brown dwarfs?

There are two criteria that divide the two categories. Brown dwarfs shine briefly by the fusion of their small amounts of deuterium into helium. But as mass and temperature drop, so does the internal temperature and even that limited ability. At a limit of about 13 Jupiter masses – just under 0.013 solar masses – even deuterium burning (fusion of deuterium into helium) must halt, and perhaps it would be fair to call the body a planet. Perhaps more significant is the way the bodies are formed. Stars and brown dwarfs are expected to be created whole from the dusty gases of interstellar space, whereas planets are assembled from smaller pieces, ultimately from the leftover dust and gas that orbit in a disk about a fully formed star.

The radial velocity technique has become accurate enough to be able to find bodies with masses in the range of the giant planets of our own Solar System, of those of Jupiter and Saturn. So far, nearly 50 of them have been discovered this way that have lower-limit masses below the 13 Jupiter-mass limit, beneath what we would normally consider a brown dwarf. Some are very strange: we see Jupiter-like "planets" tucked up next to their stars closer than Mercury would be were they in our own Solar System. Are we uncovering different styles of planetary system, or different styles of brown dwarfs? Do the criteria overlap? Can planets be created from dust that are massive enough so that they can fuse deuterium and thus behave like brown dwarfs? Can planet-like bodies be made whole, like stars? Or are there other kinds of bodies unlike either what we know of brown dwarfs or planets? We do not know.

The dim lower-mass M and L stars may help lead the way toward planetary discovery. Having smaller masses themselves, they are more easily deflected by even lower-mass orbiting companions. The M2 dwarf Lalande 21185 may move back and forth in position, in its proper-motion path, deflected by not just one but possibly by two bodies with masses about equal to that of Jupiter, orbiting with periods of 5 and 30 years, not unlike those of Jupiter and Saturn.

Whatever the outcome of these studies, we are witnessing the opening of a new science of extrasolar planetary astronomy, one that will lead us down new pathways of discovery, led by the faintest and coolest of stars, by the dim and cool M and L dwarfs at the bottom of the main sequence.

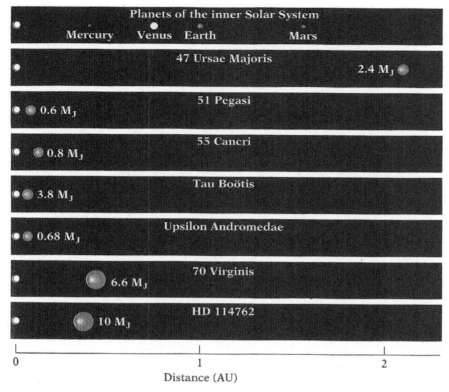

Figure 2.15. A selection of newly discovered planets shows that our Solar System is neither typical nor unique. Several Jupiter-like planets huddle next to their stars (the planets are exaggerated in size for clarity). ($M_J$ refers to Jupiter mass.) [From G. Marcy *et. al.*, as redrawn in *Cosmic Clouds* by J. B. Kaler, © 1997 by Scientific American Library. Used with permission of W. H. Freeman and Company.]

## Chapter 3

# The coolest stars . . . continued

Nature is rarely straightforward. In an exploration of stellar extremes, some stars will fall into more than one category, and one category can contain more than one kind of star, leading to "these stars are the [coolest, hottest, smallest . . .], except, of course, for these others." Such apparent contradictions can be used as transitions from one kind of star to another and can reveal a variety of differentiating phenomena.

The stars at the bottom of the main sequence are indeed the faintest as long as we refer to the visual part of the spectrum and to some kind of "normal" star like the Sun. Under different terms, however, highly evolved (what we might refer to as "dead") stars can compete. In the visual part of the spectrum, the faintest white dwarfs (the cooling remains of solar-type stars) are several magnitudes brighter than red dwarfs. However, the faintest red dwarfs are cooler than the faintest white dwarfs, and radiate much of their energy in the invisible infrared. When we take this "missing" energy into account, the faint limits of the two kinds of stars are comparable to each other (the red dwarfs do indeed just "win," if that is an appropriate term for the least of something).

Ultradense neutron stars (the collapsed remnants of exploded stars) also compete with red dwarfs as for faintness. The subset of "pulsars" can emit powerful bursts of radiation. But in between these apparent outbursts, pulsars and the other non-pulsing neutron stars are essentially invisible. Where we have information, their absolute visual magnitudes are about the same as the dim bulbs at the end of the main sequence. Neutron stars are extremely hot, however, and radiate most of *their* energy in the ultraviolet and X-ray parts of the spectrum. When we take this radiation into account, neutron stars (ignoring the "pulses") are comparable to the Sun in absolute brightness, and lose the fight for dimmest. Fainter yet are the infamous black holes, also the collapsed remnants of exploded stars, and bodies that in their pristine state emit no radiation at *all*. But can we really call them "stars?" Moreover, under some circumstances, the surroundings of black holes can perversely be very *bright*. When

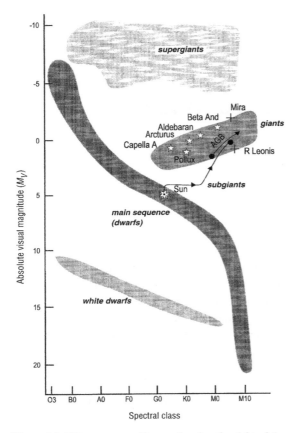

Figure 3.1. Giant stars march upward and to the right of the main sequence, the brightest 20 magnitudes more luminous than the corresponding dwarfs. Helium-burning giants are indicated by stars. The positions of two representative Mira variables at their visual brightest are indicated by crosses. The solid lines show how the Sun will evolve from the main sequence to become a giant. Helium-burning starts at the heavy dot on the lower curve. The star then settles down to the lower heavy dot. When the helium is gone, the star ascends the asymptotic giant branch (the AGB). The stars indicated here are more massive than the Sun.

all the criteria are taken into account, the red dwarfs can still fairly be called "the faintest stars." Moreover, white dwarfs and neutron stars are faint because they are so small, as are black holes, each defining its own category of "the smallest stars." Simple classification of superlatives is therefore insufficient; we need to look deeper, into the hearts of the stars themselves.

## Giants and dwarfs

As a result of their small masses, the deep-red dwarfs are also the coolest stars (at least if those in the warm end of the L class actually turn out to be real stars) with temperatures somewhat below 2000 K. And if you wish to call a brown dwarf a real "star," then we descend to cooler temperatures yet, in the one case of GL 229B to 1000 K. But the broad category of "coolest" also includes the vastly larger and evolved M giants, which at their lowest temperatures are almost as chilly as the stellar L dwarfs. More intriguingly, these cool M giants derive from the much hotter main sequence counterparts of the cool L dwarfs, the more massive dwarfs seemingly determined to match their low-mass rivals.

Like the dwarfs of the main sequence, the classic giant stars come in a variety of flavors, falling within spectral classes G through M, and in color from yellowish to red. (O, B, and A stars have their giants too, but they are not so large.) The similarity, however, ends there. The giants are much brighter than the dwarfs. It would typically take 100 million M8 dwarfs to equal the radiant output of one M8 giant. True to their names, the giants also have dimensions that are appropriate not to the Sun but to the orbits of the inner planets. Finally, the dwarfs descend in both luminosity and radius with advancing (cooling) spectral class, while the giants behave oppositely, both brightening and growing in majestic proportion as we proceed from G to M. At their largest and coolest their radii well surpass the size of Mars's orbit.

While the cooler dwarfs have minimal impact on the night sky, the luminous giants make the principal parts of many constellation patterns; half of the 40 brightest stars are giants of one sort or another. Moreover, they are not just visible, but easy to find, as the cooler giants call out who they are by their yellow-orange to reddish colors. (We need only discriminate between the giants and much brighter and larger stars, the red *supergiants*, to be examined later. These are so rare, however, that, except for first magnitude Betelgeuse in Orion and Antares in Scorpius, they make little impact on the night sky.) Several favorites are included in this group of giants: Arcturus of spring, Aldebaran of autumn, Dubhe (Alpha Ursae Majoris) in the bowl of the Big Dipper, Kochab (Beta Ursae Minoris) positioned analogously in the Little Dipper, Pollux (Beta Geminorum), Beta Andromedae, some stars in the Hyades, and a focus of this chapter, great class M Mira (Omicron Ceti).

## Toward the coolest stars

Except for the dwarfs of the main sequence – which contain both the faintest and some of the brightest stars – the story of stellar extrema can be told through the processes of stellar evolution, the stellar aging process. We now leave the red dwarfs entirely behind, as they live so long that none is yet close to evolving into anything else. The coolest stars, the extreme giants, develop from solar-type and heavier stars up to about 10 times the mass of the Sun.

For some 90 percent of their lives, stars occupy fairly stable graphical positions on the HR diagram. Trouble – and for us, the fun – begins when a dwarf runs out of its internal hydrogen fuel. At that point the stellar core, the region in the star where the nuclear reactions had taken place, no longer has the energy to support the great weight of the overlying layers and begins rapidly (on an astronomical time-scale of millions of years) to contract. At the

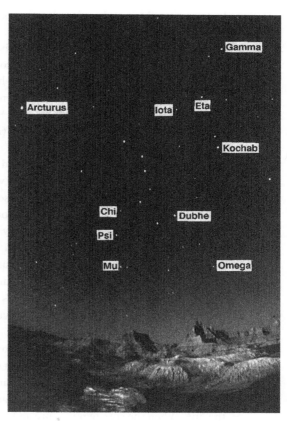

Figure 3.2. The northern sky presents several giants. The region around Ursa Major, seen descending the northwestern sky, includes Arcturus in Boötes, Chi, Psi, Mu, and Omega Ursae Majoris (as well as Dubhe), Kochab in Ursa Minor, and Iota, Eta and Gamma Draconis. [Rick Olson.]

53

Figure 3.3. (See also Plate III.) On the left is a peaceful sunset that might, when the Sun is brightening on its first ascent of the giant branch, look something like the scene on the right. [*Left*: Author's photo; *right*: from *Stars* by J. B. Kaler © 1992 by Scientific American Library. Used with permission of W. H. Freeman and Company.]

same time, the increased core temperature also heats the inner part of the old inert hydrogen envelope, and nuclear reactions spread out into a shell that surrounds the burnt-out core.

Stars therefore behave perversely. When the internal energy source gives out, they get hotter inside rather than cooler; so rather than becoming dimmer, they get brighter because of the conversion of long-held-back gravity into heat. In a mere five billion years, the Sun will begin to brighten by up to a factor of 1000 or so, its absolute magnitude climbing from its current $+5$ to $-3$, allowing our star to be a prominent part of someone else's constellation (if there is someone else), even if they be hundreds of light-years away.

During the brightening, the increased core temperature and luminosity drive the envelope surrounding the old core outward. If the future residents of the Solar System could speed up time, they would see the Sun swell in size, first becoming a subgiant and then a giant. It will pass the orbit of Mercury, and then approach that of Venus, reducing the planetary inventory of the Solar system by over ten percent.

The luminosity of a star depends on both radius (squared) and temperature (raised to the fourth power). Though the luminosity of a developing giant is increasing, its radius increases so fast that the surface temperature drops, and the future Sun will gradually lower its spectral class from G to K and then to warm M, to perhaps M2 or M3, the temperature descending to about 3200 K; and a true red giant is born.

Obviously this progression cannot go on forever. The star – the Sun in this example – does not have an infinite amount of gravitational energy available. Hidden in the atom, however, is a surprise. The current Sun is stabilized by the conversion of hydrogen into helium. But this star-manufactured element is hardly the end of the line. When the core temperature and density climb high enough, the ash of the old reaction becomes fuel for a new reaction, and the helium atoms now fuse into carbon. They must, however, first overcome an enormous barrier. Two normal helium atoms ($^4$He, with two neutrons and two protons) adding together make a highly unstable isotope of beryllium ($^8$Be), which immediately falls apart – typically

*Representative cool red giants*

The stars are arranged according to advancing spectral class (average for variables), or decreasing temperature (which changes with variation). "Period" refers to the period of variation for variable stars. The distances are from *The Hipparcos and Tycho Catalogues*. The absolute magnitudes for Mira-type variables are for visual maximum and correspond to a warmer spectral class.

| Star | Spectral class | Temp (K) | (V) | Period (days) | Variable class[a] | Distance (l.y.) | ($M_V$) |
|---|---|---|---|---|---|---|---|
| Beta Andromeda (Mirach) | M0 III[b] | 3600 | 2.06 | . . . | . . . | 200 | −1.87 |
| Mu Ursae Majoris | M0 III | 3600 | 3.05 | . . . | . . . | 250 | −1.36 |
| Delta Virginis | M3 III | 3200 | 3.38 | . . . | . . . | 200 | −0.58 |
| Mu Geminorum | M3 III | 3200 | 2.88 | . . . | . . . | 230 | −1.38 |
| R Lyrae | M5 III | 2800 | 3.9–5.0 | 46 | SRb | 350 | −0.7 |
| 19 (TX) Piscium | C5 II | 3500? | 5.5–6.0 | . . . | Lb | 760 | −1.1 |
| 30 Herculis | M6 III | 2700 | 4.7–6.0 | 70 | SRb | 360 | 0.1 |
| Chi Cygni | S6e III | . . . | 3.3–14.2 | 407 | Mira | 350 | −1.8 |
| R Leporis | C7e III | 3000? | 5.5–10.5 | 433 | Mira | 820 | −1.5 |
| R Aquarii[c] | M7 IIIe | 2500 | 5.8–11.5 | 386 | Mira | 640 | −0.7 |
| Omicron Ceti (Mira) | M7 IIIe | 2300 | 3[d]–10 | 331 | Mira | 420 | −2.5 |
| R Cassiopeiae | M7 IIIe | 2000? | 5.5–13.0 | 350 | Mira | 350 | 0.4 |
| R Leonis | M8 IIIe | 2000? | 4.4–11.3 | 372 | Mira | 325 | −0.6 |

*Notes:*

[a] SR: semi-regular; L: irregular; b is an old designation for "giant"

[b] Roman numeral III denotes that a star is a giant, II that it is a bright giant

[c] Symbiotic double star (see Chapter 7); giant component is Mira variable star

[d] Has reached first magnitude

within $10^{-16}$ seconds – back into two helium atoms. You cannot buy $^8$Be; the silvery (and dangerous) stuff commercially available is $^9$Be, which helium cannot make. So the solar engine must bypass that stage, doing so by ramming *three* helium atoms together nearly simultaneously to create common carbon ($^{12}$C). The required temperature is some 100 million kelvin within densities that approach a metric ton per cubic centimeter, 100,000 times that of lead. Another attacking helium nucleus can subsequently turn a few carbon atoms into oxygen (in the form of $^{16}$O, the most popular kind).

The star suddenly feels a new surge of youth. The unleashed power of helium-burning not only brings to a halt the core's gravitational contraction, but actually makes it expand somewhat. The newly stabilized star now reverses its direction on the HR diagram, the outer envelope shrinking (the core and envelope behaving

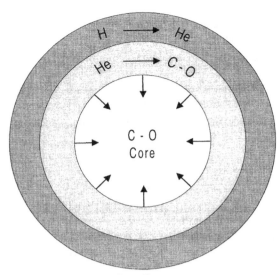

**Hydrogen**

**Envelope**

Figure 3.4. When a dying star ascends the giant branch for the second time it has a dead carbon–oxygen core that is surrounded by a shell that burns helium into carbon and oxygen, which is in turn surrounded by another burning hydrogen into helium. The two shells are not active at the same time, but switch on and off in sequence. The whole affair is enclosed in a vast hydrogen envelope thousands of times larger than the core.

oppositely). The surface heats some and becomes less red, which sends the star into the realm of the orange K giants where – in the case of the Sun – it will survive for another billion years, appearing like a dimmer version of Aldebaran or Arcturus with an absolute magnitude hovering around +1. More-massive dwarfs will occupy other regions of the giant branch, perhaps appearing more like the brighter component of yellowish, class G Capella, the Alpha star of northern winter's Auriga (the dimmer G component is making the transition to gianthood). We know that stars behave this way because clusters that have lost the upper parts of their main sequences also contain giant (or in the case of very young clusters, supergiant) stars. Theory, in fact, exactly matches what we see and provides the correct explanation.

The stability of our future K-giant Sun is again threatened when the helium runs out, leaving the core as a tiny, hot, dense ball of carbon and oxygen. The someday-Sun and its kind are now set to launch themselves into the real realm of the coolest stars. When the helium is gone, the dying star again loses its interior support, and the core resumes its contraction under gravity's inexorable force and approaches the size of Earth. As it does, the helium-burning nuclear flame, now quenched in the center, moves outward to encompass the fresh helium that was being sent inward by the hydrogen-burning shell. The evolving giant now has not one but *two* nuclear-burning shells, the inner one able to fuse helium to more carbon and oxygen, the outer one hydrogen to helium. They work in sequence, the outer one extinguishing, the inner one turning on with explosive violence as the star climbs the giant branch. As a result of all this activity, the star begins again to brighten and expand, sending it once more into the realm of the M giants with its earlier characteristics considerably exaggerated. The second time the star climbs the giant branch, it has a graphical track on the HR diagram that is crudely asymptotic to the first; these stars therefore received the peculiar name "asymptotic giant branch," or "AGB" stars. These are the focus of our attention.

The second climb up the giant branch is much faster than the first, and as these

Figure 3.5. Mira (arrow), in Cetus, is the sky's brightest asymptotic giant branch star. It is a "long-period variable" star that would easily encompass the orbits of all the inner planets. The head of Cetus is to the left of Mira. The bright star that marks its tail, Deneb Kaitos (another K giant), is near the right-hand edge at center. [Author's photograph.]

stars make their run for the top (the aftermath to be explored in the next chapter), they become much cooler than before, reaching spectral classes down to M8 or more and becoming considerably more luminous. Mira (Omicron Ceti), 420 light-years away, is the heaven's archetypal AGB star. Over 500 times the size of the Sun, its diameter is half again that of Mars's orbit, and its luminosity (including the invisible infrared radiation) is over 15,000 times solar. Even optically, the star (which is variable) could appear as much as 1000 times brighter than the Sun. To orbit such a star at a comfortable planetary temperature, our Earth would have to be more than 100 times farther away than it now is from the Sun, at a distance well over twice that of Pluto. Even from there, Mira would still be over 2.5° across, five times the angular diameter of the G2 dwarf Sun in our own sky. Near their limits, AGB stars can reach luminosities approaching 100,000 times solar.

How do we know the natures of such stars? Like those of the red dwarfs, the spectra of the cool AGB stars are enormously complex and are filled with lines and bands of molecules, chief among them titanium oxide; there are so many lines that the background continuous spectrum of the warmer gases within is all but obliterated. Several other molecules, such as vanadium oxide (VO), magnesium hydride (MgH), and even molecular hydrogen ($H_2$) and water develop at the cooler temperatures. Because of this forest of absorptions, which obscure the background continuous spectrum, it is difficult to measure the stars' surface (photospheric)

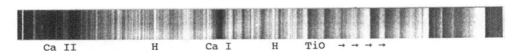

Ca II         H     Ca I     H    TiO → → → →

Figure 3.6. R Leonis, a typical asymptotic giant branch star and Mira variable, has a complex cool molecular spectrum (seen here in the violet and blue from around 3900 Å to 5000 Å) dominated by titanium oxide bands. The hydrogen lines are seen as bright rather than dark, indicating excitation by a shock wave passing through the star's atmosphere. [From *An Atlas of Representative Stellar Spectra*, Y. Yamashita, K. Nariai, and Y. Norimoto, University of Tokyo Press, 1978.]

temperatures, which are derived from theoretical modeling of the absorption spectrum. But it is from such temperatures, coupled with the luminosities, that we commonly estimate the stellar radii, rendering the radii uncertain too.

There is another approach to both size and temperature. The Sun is seen as a disk. To find the angular diameter of any object, you need only divide its physical diameter by its distance; conversely, to find a physical diameter, merely multiply the distance by the angular diameter (in appropriate angular units). Measurements of the Sun's angular diameter (32 minutes of arc, about the same as that of the Moon) and its distance yield its physical diameter of 1.5 million kilometers. All we need therefore do to find the physical diameter of a star whose distance we know is to see it as a disk and assess its angular size. Instead of deriving luminosity from temperature and radius, or radius from luminosity and temperature, we can reverse the procedures to infer temperature from the luminosity and radius; that is, we ask what blackbody temperature it would take to make a glowing body the size of the star be at the observed absolute brightness.

The problem is formidable. Stars are so far away that their disks appear very small, only a tiny fraction of a second of arc across. As a result, smearing by the telescope optics and twinkling caused by refraction in the Earth's atmosphere completely obscure the stellar disks in even the largest of ground-based telescopes and, to the eye, stars might as well be perfect points. We must find other approaches. One, rather obvious but very expensive, is to observe from space where the Earth's atmosphere does not get in the way. This technique has been successfully used with the Hubble Space Telescope to image both Mira and the supergiant star Betelgeuse.

Direct observation, however, is limited to the angularly largest of stars. A much broader approach goes back to the early part of the twentieth century and to A. A. Michelson, who built the first stellar "interferometer." Light can behave as a wave, and waves can interfere with one another. If two waves blend with each other, the crests of one can add to the crests of the other, making a bigger wave, or the waves can fill each other in, leaving only a ripple behind. If, for example, we combine light waves that emerge from two small holes drilled in a plate, together they produce a set of light and dark "fringes" where the waves alternately add and cancel one another.

Michelson's device consisted of a 10-foot (3-meter) steel beam placed across the top of the Mt. Wilson 100-inch (2.3-meter) telescope, on which was mounted two movable and two fixed mirrors. Parallel rays of light from a very distant star – one that would be a near-perfect point – would bounce off the mirrors and combine at the focus to produce such an interference pattern. But a large, nearby star with a significant disk contributes light from its entire surface. Light from opposite edges of the disk arrive at slightly different angles, and the fringe patterns produced by these opposite sides are shifted relative to each other. It is possible to find a mirror separation at which the fringes fill each other in and disappear, which gives us the stellar diameter. From this technique, Michelson, J. A. Andersson, and F. G. Pease measured the angular sizes of the supergiants Betelgeuse and Antares nearly 80 years ago to be some few hundredths of a second of arc, only a millionth or so that of the Sun.

Interferometry – the science of using the interfering properties of electromagnetic waves – has become increasingly sophisticated, and there are a great variety of different instruments in use that operate from the radio (where it is most familiar) through the infrared into the optical. Some optical "interferometers" use separate telescopes from which the individual light beams are combined, while others mix the light signals that have fallen in various isolated zones of the telescope's primary mirror. However it is done, modern optical devices are capable of measuring angular diameters with an accuracy of a thousandth of a second of arc or better, giving them the potential ability to separate car headlights on the Moon.

There are severe problems, however, in the analysis of these great stars, whether by direct observation or by interferometry. The Sun presents us with a sharp, well-defined limb. Even though darkened as a result of the temperature variations within the semi-transparent gas, we can easily measure a definitive solar diameter. But because the big class–M giants are so much less dense than the Sun their "edges" are much more diffuse, and limb darkening is much more severe. As a result, we cannot be sure just where the star's "surface" (such as it is) actually ends, and we have to define radius by a measure of how far within the outer gases we could look. Moreover, since

Figure 3.7. Mira, imaged directly by the Hubble Space Telescope, is not round. The star might be affected by a nearby white dwarf companion, but non-sphericity might also be a natural state of AGB stars, caused by instability and huge size. [M. Karovska *et al.* (Center for Astrophysics), STScI, and NASA.]

the transparency of the gases depends on the wavelength of observation, so does the diameter. Mira's angular diameter, as measured by the Hubble, is 0.035 seconds of arc in the ultraviolet, but 0.056 seconds in the visual where absorption by titanium oxide prevents us from seeing very far into the star, making it look bigger. At 420 light-years, these measurements correspond to respective diameters of 4.6 and 7.2 AU, the latter considerably larger than expressed earlier. In the middle of the visual spectrum, Mira is 2.3 times the size of Mars's orbit, and if placed at the Sun would extend two-thirds of the way to Jupiter! Worse, most of these very cool stars are variable in brightness, the variation caused by changes in both temperature and diameter. As a result, diameter measurements at different times can differ by significant factors. Worst, the stars are apparently not even round! Interferometer images of Mira variables, including Mira and R Cassiopeiae, plus a Hubble Space Telescope image of Mira, show the stars to be as much as 30% larger in one direction than in another.

What is the limit to such stars? Guidance instruments aboard the Hubble can act as an interferometer as well. With it astronomers have set angular diameter records with R Leonis and W Hydrae of 0.074 and 0.084 seconds of arc respectively. At respective distances of 325 and 375 light-years, these angular diameters transfer to physical radii of 3.7 and 4.0 AU, the latter 80% the size of Jupiter's orbit.

At their most extreme, the surface temperatures of the AGB stars dip toward those of the coolest red dwarfs. Mira's (which is variable) seems to fall just below 2500 K. S Pegasi's average temperature is measured even below that, closer to 2200 K, and the temperatures of R Leonis and R Cassiopeiae may descend as low as 2000 K similar to the coolest M dwarf. Whatever the actual values for any of these stars – and they are difficult to define – we are truly still in the realm of the coolest stars.

The correlation between temperature and spectral class is not the same for different kinds of stars. Spectral class is defined by the balance of ionization in the gas, and for the lower temperatures by the relative number of atoms that link into molecules. The abundance of molecules depends on the temperature, density, and transparency of the gas. If you raise the temperature, collisions knock molecules back into single atoms. But if you raise the density, the atoms get closer together, which oppositely promotes molecule formation. Giants are much larger than dwarfs, and therefore have much lower atmospheric densities, which act against the creation of molecules. As a result of the enormous differences between them, giants and dwarfs have different temperature scales relative to their classes. Moreover, there are no class L giants. All cool giants are class M, as defined by the titanium oxide molecule. Class L is confined to the dim versions of the coolest stars, the dwarfs.

## Carbon stars

The "red giants" are not really all that red; the term is used more to indicate relatively chilly temperatures rather than color. A close look shows the color to be nearer to orange, unless they are seen next to a white star, in which case the eye fools us into seeing them redder than they really are. Sprinkled here and there over the sky, though, are a few dim, more colorful and truly red stars. Casual observers are unaware of them because even the brightest is only fifth magnitude, their faintness rendering them colorless to the naked eye. The telescope or photography, however, shows their true natures.

These ruby-like stars clearly have low temperatures like those of class M, but their spectra are entirely different. Even in the nineteenth century, the pioneer spectroscopist Father Angelo Secchi noted that the stars have molecular bands that fade in the direction opposite to those found in the more normal M stars. The Harvard astronomers of the 1890s placed them not into class M (which requires TiO bands), but into a neighboring class N. The molecular absorptions are caused not by metallic oxides, but are produced instead by molecules of carbon, notably $C_2$, CN, CH, even carbon monoxide and carbon dioxide. TiO is nowhere in sight. These are the "carbon stars."

Carbon stars also extend to warmer temperatures like those of class K, where they are called R stars. The R and N classes are nowadays collectively referred to simply as type C (for carbon). The red color of the cooler N stars is a result of an enormous amount of absorption by molecules and neutral metals that nearly eliminate the blue and violet portions of the spectrum. The spectra are so complex that we have considerable difficulty evaluating the stars' temperatures, but we know that

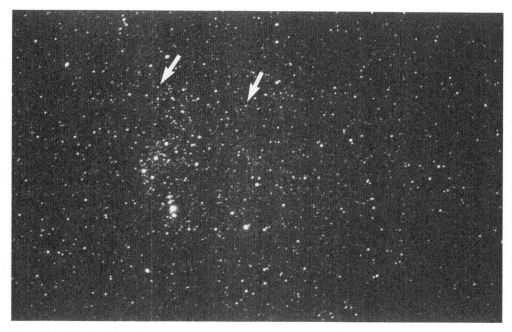

Figure 3.8. W Orionis, a low-amplitude semi-regular (class SRb) variable (*right-hand arrow*), is one of the brightest carbon stars in the sky. To the east of it (*left-hand arrow*) is CK Orionis, a bright irregular or semi-regular class K oxygen-rich variable. [Author's photograph.]

$\leftarrow \ \leftarrow \ \leftarrow$ CN　　　CH　　　$C_2$　　　$\leftarrow$CN$\rightarrow$　　　$C_2$

Figure 3.9. The blue-violet optical spectrum of the carbon star 19 Piscium, the brightest carbon star, is dominated by bands of the $C_2$ molecule; TiO is absent. Barium, an s-process element that is made by nuclear processes inside stars, is quite prominent. Also known as TX Piscium, the star is a low-amplitude irregular or semi-regular variable. [From *An Atlas of Representative Stellar Spectra*, Y. Yamashita, K. Nariai, and Y. Norimoto, University of Tokyo Press, 1978.]

they extend downward toward 2000 K and are, on the whole, seen to be a bit lower than those of class M, perhaps even invading the realm of the L dwarfs.

The distinction between carbon stars and the more ordinary AGB giants lies in the ratio of the number of atoms of carbon to those of oxygen, that is, in the C/O ratio. Carbon and oxygen have a great affinity for each other, and if there is less C than O, there will be enough oxygen left over to form the metallic oxides, whose absorbing power lets them dominate the spectra. But if there is more carbon than oxygen, all the oxygen is used up in carbon compounds, and the excess carbon simply combines with itself.

On the cosmic scale, oxygen is about twice as abundant as carbon, and M stars are far more common than C stars. But in the C stars C/O must be greater than 1 (how much greater is not well known, as there are no oxygen absorptions that do not involve carbon). How does such a change happen? We are (in a broad sense) actually seeing stars altering their compositions as they age. The transformation begins on the AGB and seems to affect only stars two or three times as massive as the Sun. The helium that burns in the shell around the core leaves a residue of carbon. In the outer parts of the star, energy can be transported outward by convection: hot gases from below rise, release their energy, cool, then fall. Under particular circumstances that are still not fully understood (convection being one of the more difficult problems in astrophysics), the convection currents can dip deep enough into the star to meet other currents coming up from the carbon-rich zones below, the combination cycling the element to the surface. Most of the carbon on Earth, including that with which you and I are made, was produced by this mechanism.

Since the process is gradual and also depends on mass, we would expect to see intermediate cases. Between the M and C stars lies class S, for which we now know that C/O is equal to 1. The S stars, which dip to temperatures as low as the M and C stars, were originally recognized by *zirconium* oxide in their spectra, the ZrO replacing the TiO of the M stars. Zirconium has a greater affinity for oxygen than does titanium, so when carbon is about equal to oxygen in quantity, the small amount of oxygen is preferentially used up by the zirconium, leaving none for the titanium.

Carbon is not the only element manufactured and dredged to the stellar surface. Reactions deep within the nuclear-burning zone release neutrons that can attach themselves to lighter elements to build heavier ones through the "s-process," shorthand for "slow neutron capture." The atoms do not capture slow neutrons, they

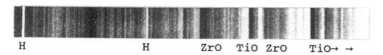

H                H        ZrO  TiO ZrO    TiO→ →

Figure 3.10. Chi Cygni, another long-period Mira variable, is the sky's brightest S star, the class easily recognizable by violet and blue bands of zirconium oxide mixed in with titanium oxide. Bright hydrogen emission lines (the left two at 4101 Å and 4340 Å) indicate shock waves induced by Mira pulsation. [From *An Atlas of Representative Stellar Spectra*, Y. Yamashita, K. Nariai, and Y. Norimoto, University of Tokyo Press, 1978.]

capture them slowly, one at a time. The process is responsible for a portion of most of the elements heavier than iron, including almost all the strontium, zirconium, and barium in the Galaxy – and in the Earth.

All atoms have a wide range of isotopes, in the number of neutrons that can be attached to the nuclei, and therefore in atomic weights. But if there are too many neutrons (or, for that matter, too few), the atom becomes unstable and in some manner or other changes form. One common way is by "beta decay," in which a neutron inside an unstable atom kicks out a negative electron (which was at one time called a "beta particle"). To balance the removal of the negative charge, the neutron must take on a positive charge and become a proton, which increases the atomic number, that is, the atom moves up a notch in the periodic table.

Look at how the zirconium is made. The element below zirconium (number 40, that is, 40 protons) is yttrium (Y), atomic number 39, which has but one stable isotope, $^{89}$Y. It can absorb a neutron to become $^{90}$Y, which is so unstable that it "beta decays" into a stable isotope of zirconium ($^{90}$Zr). But some of the $^{90}$Y can stick around long enough to absorb another neutron, which then decays to stable $^{91}$Zr. At the same time, more $^{91}$Zr can be made by the absorption of a neutron by $^{90}$Zr. The process is limited only when the number of neutrons captured onto the nucleus gets so large that the isotope decays almost instantly, leaving none for neutron-absorption. The s-process consists of an enormously complex network of reactions that can take elements from iron to bismuth. Since zirconium is a strong link in the s-process, the process adds powerfully to the S-star effect, in which zirconium oxide replaces titanium oxide as carbon (and the various "s-process elements") rise to the star's surface.

Direct proof that we are witnessing the actual creation of elements comes from the observation of the element technetium (Tc) in cool giants. This substance, number 43 in the periodic table, is highly radioactive, and decays very quickly into something else with the release of radiation. Like most elements, it has a variety of isotopes, all of them unstable. The longest-lived of them, $^{97}$Tc, has a half-life (the time it takes for a radioactive element to decay to half its original amount) of only two million years, far less than the lifetime of a star, and is effectively absent from Earth. If it were not freshly manufactured by the s-process acting on molybdenum it could not possibly be seen, vividly showing that stars are dynamic affairs that bubble with energy from the inside all the way out. More important, observations of

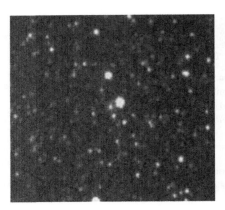

Figure 3.11. The long-period (Mira) variable star TT Monocerotis (the left-hand star of the pair in the center) shows dramatic changes from 12th magnitude (*left*) to 8th (*right*) over a period of about 5 months. [University of Illinois Stardial system.]

these newly cooked elements allow us a "look" of sorts into deep stellar interiors, in which we infer processes and conditions by theory from what we see on the stellar skin.

## Variability

In late northern autumn evenings, as the warmth of the summer skies relinquishes itself to a decided chill, watch great Cetus cross the heavens, the Sea Monster ever stalking Andromeda. The figure is not all that well known, even though ancient, as there are few bright stars. You may recognize Alpha Ceti – Menkar, "the nose" – just under second magnitude, or more likely second magnitude Beta – the "whale's tail" (Deneb Kaitos) – set into a blank region of the southern sky northeast of Fomalhaut. If you watch Cetus over a long period, however, you may see yet a third prominent star. Most of the time it is quite invisible, faint and below the level of human vision. But on occasion, roughly once a year, Mira, the "wondrous star" and the archetype of the AGB stars, may reach Alpha's stature, or perhaps rival Beta. Known to us for centuries, the star typically changes from second magnitude to tenth and back over a period of 335 days.

Just as variability is common for the cool M dwarfs, so it is for cool class M, C, and S-type AGB giants, such stars known collectively as "Mira variables" after their grand prototype. Thousands of Mira variables are known. The mechanism of variation is, however, completely different than it is for the M dwarfs, which are flare stars, their impulsive variations caused by the sudden release of stellar magnetic fields. The Mira giants vary by pulsation, by continuously changing their radii, temperatures, and spectral classes.

The prototypical pulsators are the Cepheids. Cepheids are F and G supergiants that occupy a nearly vertical "instability strip" on the HR diagram. The variations

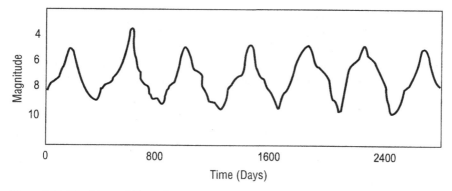

Figure 3.12. The M-type Mira R Hydrae varies between 5th and 10th magnitude with considerable, but not perfect, regularity over an interval of about 346 days. [From *Burnham's Celestial Handbook*, R. Burnham Jr, Dover, 1978.]

are produced by internal effects that cause the stars to expand and contract with remarkable regularity. The pulsations are generated by the ionization of hydrogen or helium in layers that are at just the right depth. As a Cepheid pulsates, the internal temperature and density change. The act of ionization takes energy, and any change in temperature – which affects ionization – causes the ionizing layer to act as a valve that can absorb or release heating radiation. The valve and the pulsation feed back into one another, and the pulsation continues until evolution causes the internal structure of the star to change to a more stable configuration.

Cepheids typically vary by a magnitude or two over a period of a few, or a few tens, of days. Their major claim to fame is that their periods are tightly correlated with their luminosities or absolute magnitudes. As a result, Cepheids make wonderful distance indicators. We need only measure a Cepheid's period to know its absolute magnitude; comparison with apparent magnitude then yields distance. (It has not been that easy of course. Measurement and accurate establishment of the correlation have taken most of the twentieth century.)

The cool Mira variables operate on much the same principles as do the Cepheids, but they are so large that they take much longer, commonly a year or more, to oscillate. Periods are known up to 1000 days. As a result, Miras are alternatively known as "long-period variables," or LPVs. Their large sizes, however, make them much less regular than the Cepheids, their periods varying by many days and their maximum brightnesses by a couple of magnitudes or more. They occupy their own instability zone in far right-center portion of the HR diagram, and though the LPVs exhibit a crude period–luminosity relation, it is not really good enough to establish accurate distances.

All Miras are AGB stars, but not all AGB stars are Miras. As stars begin to climb the giant branch of the HR diagram for the second time (with their carbon–oxygen cores), they are relatively stable. The big pulsations only begin late within AGB evolution, as the stars approach the tops of their tracks in the HR diagram and reach their maximum average brightnesses and radii. Since AGB stars can be of class M, C, or S, so can the Miras. The most famous Mira of class M is of course Mira itself, but several others, like R Hydrae and R Aquarii, reach naked-eye view. The best example of an S-type Mira is Chi Cygni (3rd to 14th magnitude over 406 days); the most famous carbon-star Mira is probably R Leporis in the dim constellation Lepus, below Orion (5th to 10th over 432 days).

The huge variability seen with the naked eye is deceptive. These stars are all so cool that most of their radiation emerges in the infrared where the eye cannot see it. Like their dimmer dwarf cousins, Miras are actually two to four magnitudes more luminous than we actually perceive them. Absolute visual magnitudes reach perhaps to $-4$, but a magnitude scale based on total radiation might yield $-7$, corresponding to a star a hundred thousand times more luminous than the Sun. In the infrared, Mira itself varies by only a magnitude or two. But as the star pulses, the surface temperature changes and the spectral class goes from M5 to M9. When the star cools, the bulk of the radiation shifts even further into the infrared, sharply diminishing the amount of visible light. Also important, decreasing temperature promotes the formation of yet more TiO molecules, their powerful absorption bands reducing the optical brightness even more.

Because a Mira's temperature changes along with its pulsing radius and increases with depth into the star, and because the transparency of the stellar gases changes with wavelength (so that at different wavelengths we see to different temperature levels), the optical and infrared maxima do not take place at the same time, the optical maximum coming first by up to two months. It is hard to imagine what these stars actually look like. Since we know they are not even spherical, their oscillations are probably not symmetric either. One can imagine an oscillating blob something like a balloon filled with water tossed in the air.

Other kinds of variables, the "semi-regulars" (SR), and the "irregulars" (L), fill the realm of the cool stars as well. The semi-regulars are subdivided into SRa, SRb, and SRc. The last category refers to variable supergiants, and we will look at them later. The SRa stars (S Aquilae, R Ursae Minoris, and V Hydrae, the last a carbon star) are similar to Miras but exhibit smaller light variations and irregularities in their periods. The SRb types (4th magnitude R Lyrae, W Orionis, R Crateris, and U Hydrae) have periods that are less well defined. The irregulars, subdivided Lb for giants and Lc for supergiants, have no discernible periods at all – we see simply erratic wanderings.

The evolutionary relationships among the irregulars, semi-regulars, and the Miras is not entirely clear. Different types concentrate into different spectral classes: oxygen-rich Miras spread over all M classes, whereas carbon-rich Miras concentrate toward the cooler categories; SR and irregular variables concentrate strongly toward the carbon stars, and into classes M6 and M7 in the oxygen-rich stars. The SR variables, however, seem to be essentially Miras with smaller light variations and shorter periods. The irregulars might be trying to be Miras but cannot do it. A Mira might be loosely likened to a violin string. It has a fundamental audio frequency or tone,

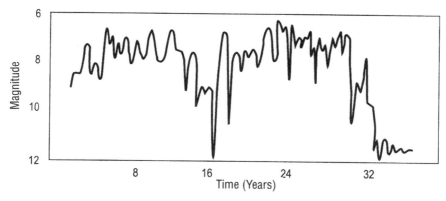

Figure 3.13. The semi-regular carbon-star variable V Hydrae (one of the visually reddest stars in the sky) fluctuates erratically with a crude 530-day period. [From *Burnham's Celestial Handbook*, R. Burnham Jr, Dover, 1978.]

which gives it pitch, but at the same time oscillates at higher frequencies, or overtones, which provide its unique audio character. Some Miras – especially those with longer periods – pulsate in their natural fundamental frequencies. Others seem to oscillate in the first overtone (a subject that can produce a good argument at a meeting of variable-star astronomers). The giant (Lb) irregulars may just be beginning their ascent of the AGB, may be oscillating in several overtones at the same time, and have yet to find their way in the world.

## Mass loss

Mira "atmospheres" (the gaseous surfaces in which their spectra are produced) present scenes of considerable violence. The stars are much too cool to allow absorption lines of hydrogen. Instead, we observe bright hydrogen *emission* lines. Such lines are commonly produced by high-energy radiation that has elevated atomic electrons that subsequently fall back down, losing their energies and emitting photons. But the Miras' radiation energy is too low, so the emissions have to be excited by some other means.

Instead, the emissions show us the existence of great outwardly moving shock waves that are apparently produced by the pulsations. Shock waves develop when matter moves faster than the speed of sound. The best analogy is a sonic boom from a supersonic aircraft, which violently pushes air out of the way faster than it wants to move. The result is a sudden wall of pressure extending from the source that hammers anything it hits (hence the loud "boom"). Because of the emissions, the letter *e* is appended to the spectral class when they are present.

Except for cooled white dwarfs (and black holes, if we dare call such things "stars"), all stars lose mass through winds (even hot white dwarfs and neutron stars). The solar wind, at least in part magnetically activated, removes a ten-trillionth of a solar mass from the Sun every year: enough to make terrestrial auroras as the wind hits us, but not enough to affect the Sun very much. Giants, however, have much higher luminosities and lower surface gravities than the Sun, and the rates of mass loss rise to much higher levels, achieving their apices in the advanced variable AGB stars, stars that can be larger than the inner Solar System.

These stars, which have been losing mass all along, now begin to evaporate under the onslaught of the atmospheric shock waves. Some of the matter lofted upward condenses into tiny grains of dust smaller than anything you could see. The dust is pushed outward by the pressure of light coming from the enormously luminous star. We see the same effect in a comet's tail, the dust released from the comet's fragile nucleus repelled backward by the action of sunlight. As the dust moves outward it drags the gas along with it, and the whole assembly flows as a powerfully expanding sphere of dirty gas. Mira loses matter at a rate of about $10^{-7}$ solar masses a year, the higher-mass versions at rates hundreds of times greater. While even this number may seem small, note that the star can be in this state for tens of thousands of years. The wind flows may also be massively enhanced by the occasional, sudden, explosive turn-on of the inner helium-rich shell, which sends a pulse of energy through the star. Over its whole giant evolution, the Sun will lose nearly half of itself back into space as it begins to expose the fiery inner core that will someday be seen as a dense, compacted white dwarf. More-massive stars can lose 80% or more of their original substance.

As matter flows outward, a massive Mira can enclose itself in a cocoon of its own making, its dust particles absorbing so much starlight that the expanding cloud hides the inner star from view, at least in the optical part of the spectrum. But the absorbed stellar radiation heats the dust grains to a few hundred degrees kelvin, and in the infrared the shroud may glow brightly, revealing the celestial location of the mysterious star. The infrared satellite observatory *IRAS*, which orbited the Earth in 1982, found great numbers of such stars. They can now be studied in detail by more advanced satellite systems. The natures of the buried stars are readily revealed by the spectral signatures of the outflowing dusty envelopes. In the infrared we may see broad emissions from silicate grains, made of oxygen and silicon, characteristic of an M-type oxygen-rich Mira. Or we may see radiation from silicon carbide, which shows us that a carbon star is buried within.

The chilly conditions of the expanding dusty envelopes favor the production of molecules whose emissions can be detected in the radio spectrum. M-star winds emit spectrum lines of silicon monoxide and often powerful emission

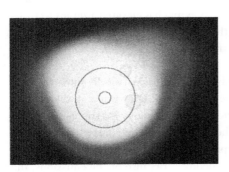

Figure 3.14. (See also Plate IV.) The carbon star IRC+10216 is surrounded by a dust shell of its own making. The circles indicate the position of the star and the inner edge of the dust shell. [S. Ridgeway and J. Christou, AURA/NOAO/NSF.]

lines of hydroxyl, OH. These extreme evolved infrared(IR)-bright Miras are thus termed "OH/IR stars." The OH lines radiate as natural masers from the outer parts of the stellar shroud at distances of some 500 AU from the star. The maser is the microwave version of the familiar laser, an acronym for *l*ight *a*mplification by the *s*timulated *e*mission of *r*adiation. A photon may have just the right energy to raise an electron from a lower orbit to an upper one, the photon being absorbed in the process. However, in a reverse process, the photon can also stimulate an electron that is already in the upper orbit to drop to the lower orbit. The original photon is not affected and is joined by a photon created by the energy-losing electron. The two photons fly off in concert, in the same direction and with their "waves" oscillating up and down together. The stimulation thus has an amplifying effect, and the sum of all stimulations produces a powerful, narrow beam of radiation.

Physical systems seek their lowest energies. In a normal gas there are always more electrons in the lower orbit of a pair than in the upper orbit, so that there is never any net amplification within the whole system of atoms: absorptions and emissions (both spontaneous and stimulated) balance each other. However, if we can tip the orbit populations upside down and put more electrons on top, the stimulated emissions may dominate, generating the laser (or maser) beam. The problem is to "pump" the gas with energy in order to raise the electrons to the proper level. In a Mira the pumping is done by the star's own infrared radiation: we can actually see the masering radio lines vary with the same period as the star.

The emission lines radiated by OH are doubled. Such a phenomenon, quite common in astronomy, is readily interpreted as being caused by emission from a shell in which the far side is expanding away from us (relative to the star) and the near side is expanding toward us. (The same thing would be caused by a contracting shell, but that would make little sense given the natures of these stars). From the degree of splitting, we find expansion velocities of 10 or 20 kilometers per second. The wind, we see, blows rather gently (compared with explosive release in which the expansion velocities can reach thousands of kilometers per second).

Since the OH maser is pumped by the star, the OH variations must follow the stellar variations; but they must be delayed by the time required for the light to get from the star to the molecular shell and then (in the form of longer radio waves) to us, and it takes longer for the radiation from the back, more distant, side of the shell to reach us than it does for that from the front side. What we thus actually observe is that the two line peaks are out of synchrony with one another by the light travel-time across the shell. All we have to do is measure the time between the maximum for one peak compared to that for the other and divide by the speed of light to find the shell's diameter, which typically is some ten thousand astronomical units, or a tenth or so of a light-year, across. At 10 kilometers per second, the wind must have been blowing for several thousand years to reach that size, which gives a rough idea of how long the star has been in this state. And with a mass loss rate of $10^{-4}$ solar masses per year, several tenths of a solar mass must be tied up within the shell's dusty confines. Finally the shells are so large that we can measure their angular diameters (several seconds of arc even at several thousand light-years distance) which, combined with their physical diameters, yields distances! These stars are a true treasury of information.

Masering does not stop with simple OH. Much closer in, perhaps 100 AU from the star, we find water masers that are pumped by collisions with other atoms, and closer yet, nearly within the extended stellar atmosphere, we find masered radiation from silicon monoxide. The dusty shells around the carbon-star Miras are in some ways even more fascinating. Here though, we see little molecular maser emission. What there is includes radiation from hydrogen cyanide and silicon sulfide. More important, carbon has the propensity to form complex molecules. In the archetypal object, known only by its catalogue name IRC+10216, we observe over 20 different kinds, including such organic species as ethylene ($C_2H_4$), methyl cyanide ($CH_3CN$), and cyanohexatryine ($HC_7N$). Such observations show the propensity of molecules to form anywhere they get a chance, and help illuminate the nature of the chemistry of the dusty gas in true interstellar space as well as that of the cloud of dusty matter out of which our Earth was formed.

Both the carbon-star Miras and the OH/IR Miras are of great significance in the evolution of the Galaxy. The outbound matter adds considerable mass – mostly hydrogen and helium – to the thinly spread matter between the stars, the so-called "interstellar medium," and salts it with carbon and other heavier elements as well.

As a result of this process, in combination with the ejecta of exploding stars (which are more important for most elements), the interstellar gas becomes ever more enriched, explaining the well-established fact that the proportion of heavy stuff is smaller in old stars than it is in young ones. Most of the carbon on Earth, including that out of which we are made, was produced aeons ago in giant stars and expelled into the clouds from which the Sun was born.

Perhaps more important, AGB stars are responsible for most of the dust in interstellar space, carbon dust (something akin to graphite) from the carbon stars, silicates from the oxygen-rich Miras. There the tiny grains accumulate ices and metal atoms, and help form the great dark clouds of the Milky Way, giving it the wondrous structure that can be seen with the naked eye on any clear dark night. The dust in the clouds

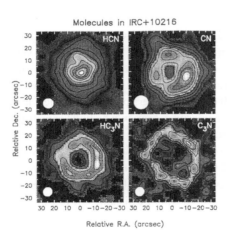

Figure 3.15. (See also Plate V.) Different molecules appear in different locations in the ejected shell around the carbon star IRC+10216, each dependent on different conditions and the degree of illumination by starlight. Each map is 6000 AU across. From inside out we see hydrogen cyanide (HCN), cyanogen (CN), cyanoacetylene ($HC_3N$), and unstable cyanoethynyl ($C_3N$). [J. Bieging and the Berkeley–Illinois–Maryland Association.]

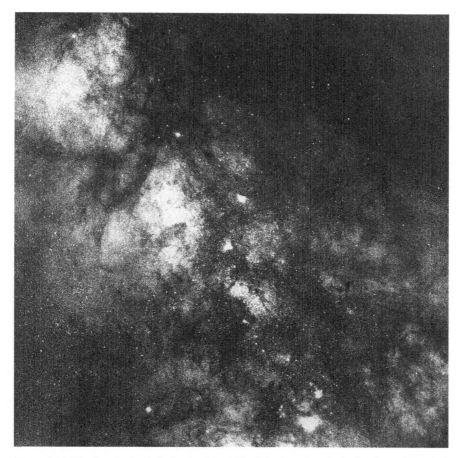

Figure 3.16. The dust in the dark clouds of the Milky Way (seen here in Sagittarius and Scutum) was produced mostly by windy asymptotic giant branch stars. The dust chills the gas and allows star formation. [From the *Atlas of the Milky Way*, F. E. Ross and M. R. Calvert, University of Chicago Press, 1934. Copyright Part I 1934 by the University of Chicago. All rights reserved. Published June 1934.]

shields the clouds' interiors from the heating effects of starlight, and they grow cold, to near absolute zero. The chilly temperatures allow the denser portions of the clouds to collapse under their own gravity. The contraction heats the interiors of the collapsing blobs. When they get hot enough, the hydrogen "fires up", the proton–proton chain turning on to make helium. The new energy stops the contraction, and a star is born.

Rotation can spin the remaining dusty gas around the outside of the star into a disk. The dust grains – some primitive, others born within – accumulate and grow, first to what we now call comets and bodies like asteroids, and then into planets, some of which – like Earth – are made of the heavy atoms produced almost entirely in stars. We can therefore think of stars as vast recycling engines that pump the bulk of their mass back into space to help create new generations, much of the action taking place in the coolest stars, the great Mira variables.

We might consider – with some sympathy – what happens to any planets that may have been orbiting such stars. From many studies astronomers are generally convinced that most stars have planetary systems and that their formation is a natural by-product of stellar formation. As a main sequence star evolves and becomes more luminous, any encircling planets must inevitably become hotter. As the developing giant, and then as the AGB star, encroaches on its inner planets, the planets may melt and finally vaporize, adding their heavy elements, silicates and carbon, into the expanding envelope.

In our Solar System, Mercury is certainly doomed, and the Sun may expand to fill the orbits of Venus and even Earth. However, the future giant Sun will be losing mass, causing these two planets to move away, possibly to save themselves, not that it will matter much as the temperature will be far in excess of that allowed for life. During the process, the heating Earth may well become more like Venus is today, with a hot (470 °C) carbon dioxide greenhouse atmosphere. On the more positive side (and it is hard to find one), the subterranean water on Mars may melt, helping give Mars a temporary thick atmosphere, making it more like Earth and perhaps livable.

The great heat will cause enormous changes in the outer planets as well, in addition to vaporizing the inner cloud of comets, those that lie just outside the orbit of Neptune. These are the primitive small bodies that might have been a planet, but there were not enough of them out there to make one. Since the comets are made of dust and ice (accumulated from interstellar space), they will contribute both solid matter and water vapor to the giant's powerful wind. As the Sun loses its mass and gravitational grip, a huge number of comets in the outer comet cloud, which extends a good fraction of the way to the nearest star, will be lost to space, to roam the spaces between the stars forever.

Yet for all the devastation they might cause, the coolest stars, representative of giant stars in general, are profoundly important to us. Without the dust and heavy elements they help to create, we – the Sun, the Earth, and ourselves – would not be here.

# Chapter 4

# The hottest stars

Over their lifetimes, stars of intermediate mass, those like the Sun, climb the giant branch twice. The first time up they end their ascent and stabilize by suddenly fusing the inert helium in their cores into a mixture of carbon and oxygen. What, however, stops the second ascent? The stars cannot continue to become ever brighter, else the night sky would no longer be dark. There must be a closed door at the top of the staircase. The shrinking carbon–oxygen core could – like the helium core before it – begin fusing into heavier elements, into a mixture of neon, magnesium, and oxygen. Such nuclear burning would stop the contraction and stabilize the star once again. However, to fuse carbon the core must get even hotter and denser than required for helium burning, conditions that can be supplied only by compression.

Such compression is not available, as the evolving star is losing its exterior hydrogen envelope, its great piston, through winds. Unless the initial star – as it was newly born – was very massive to begin with (over about 10 solar masses), the remaining thin envelope will not be heavy enough to provide the necessary internal heat and pressure. The carbon–oxygen mixture in the cores of these lower-mass stars is therefore doomed to sit there forever but, because of the enormous loss of mass, in an exposed state that can become no brighter and in which we see the coolest stars turn temporarily into one kind of the hottest stars. (The rarer neutron stars constitute another, even hotter set that will be examined under the title of the "smallest stars," again showing the overlap among stellar types.) The transformation is accompanied by an ephemeral phenomenon that could allow this chapter equally well to be called "The prettiest stars."

## The planetary nebulae

None of the faintest stars is even close to being visible to the naked eye. On the other hand, the coolest stars, represented by the red giants, are seemingly everywhere. As we turn from the coolest to the hottest stars, they again drop below naked-eye vision,

*Some planetary nebulae*

| Nebula | Common name | Constellation | Distance[a] (ly) | Radius (ly) | Apparent visual mag.[b] (V) | Temperature[c] (K) | Comments |
|---|---|---|---|---|---|---|---|
| NGC 40 | | Cepheus | 3500 | 0.30 | 10.65 | 32,000 | very low excitation |
| NGC 650-1 | | Perseus | 2400 | 0.80 | 16.30 | 135,000 | M76, large outer shell |
| NGC 2392 | Eskimo | Gemini | 4000 | 0.45 | 10.53 | 80,000 | double shell |
| NGC 2440 | | Puppis | 3600 | 0.30 | 17.66 | 220,000 | hottest confirmed |
| NGC 6543 | | Draco | 3200 | 0.15 | 11.31 | 47,000 | AGB halo visible |
| NGC 6572 | | Serpens | 1800 | 0.06 | 12.86 | 60,000 | very bright |
| NGC 6720 | Ring Nebula | Lyra | 2500 | 0.40 | 15.00 | 145,000 | M57; large outer halo |
| NGC 6826 | | Cygnus | 5000 | 0.30 | 10.69 | 47,000 | AGB halo visible |
| NGC 6853 | Dumbbell | Vulpecula | 900 | 0.75 | 13.82 | 160,000 | M27 |
| NGC 7009 | Saturn | Aquarius | 2900 | 0.20 | 11.30 | 80,000 | first discovered |
| NGC 7027 | | Cygnus | 3000 | 0.10 | 16.26 | 174,000 | big molecular cloud |
| NGC 7293 | Helix | Aquarius | 500 | 1.0 | 13.43 | 120,000 | closest planetary |
| NGC 7662 | Ring Nebula | Andromeda | 3500 | 0.25 | 13.20 | 100,000 | double shell |

*Notes:*

[a] Most distances are estimates

[b] Visual magnitude of the central star

[c] Temperature of the central star

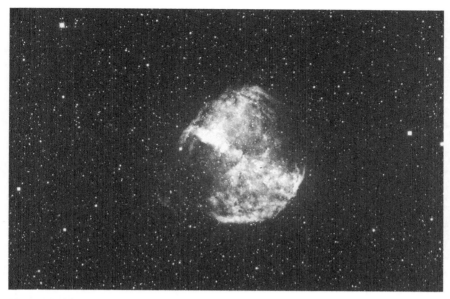

Figure 4.1. The Dumbbell Nebula, M27, in Vulpecula, appears as twin lobes of light. It is lit by ultraviolet radiation from the faint, very hot star at its center, a star that was once at the core of a Mira variable and that will eventually become a white dwarf. [AURA/NOAO/NSF.]

but not so far that these stars and the phenomena associated with them are not easily visible with but modest optical equipment.

Some northern summer evening turn your attention to Cygnus, the Swan ever flying his celestial path to the south along the Milky Way. Just east of Albireo, the star that marks Cygnus's head, locate a small W-shaped figure, the most prominent part of the obscure modern constellation Vulpecula, the Fox. Aim a pair of binoculars at the middle star, and just to the east of it you may – if the sky is good and dark – note a small fuzzy patch of light, something like a star gone soft. Turn to your telescope and be enthralled by a pair of ghostly globes of light that have an almost three-dimensional appearance against the star-spattered sky. You are looking at the "Dumbbell Nebula," a prime example of a "planetary nebula." The Dumbbell is one of the great showpieces of the sky, commonly turned to at observatory open houses where newcomers to astronomy are amazed to find that there is more in the sky than just stars. At its dead center, a larger telescope reveals a faint star.

Planetary nebulae were announced as a class of objects in 1785 by William Herschel, their disk-like appearances recalling the disks of the planets as seen through a small telescope. A musician, originally from the German state of Hannover, Herschel became far more enamored of the stars than of music, and in many ways was the founder of modern observational astronomy. His legendary discoveries include the planet Uranus, double stars, infrared radiation, a sense of the shape of the Galaxy, the planetary nebulae, huge numbers of star clusters, other kinds of nebulae, and much more.

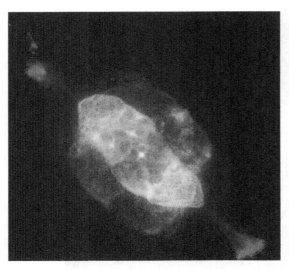

Figure 4.2. NGC 7009 is called the Saturn Nebula because of the handle-like ansae that erupt from its major axis. It was the first such object to which the name *planetary nebula* was applied. [B. Balick (University of Washington) *et al.*, STScI, and NASA.]

Planetary nebulae abound, thousands now known. At the time of Herschel's discovery, the Dumbbell was already recognized as a peculiar object. Herschel also had already found a peculiar body he called a "perforated ring of stars," the famed Ring Nebula in Lyra, which he later entered into the class of "planetaries." The term "planetary nebula" was actually first applied to the small but bright Saturn Nebula in Aquarius, the constellation that also contains the magnificent Helix Nebula, a planetary nebula whose angular extent is half that of the full Moon. Herschel was the first also to notice a central "condensation" that appeared at one of his nebulae, NGC 6543, which turned out to be a central star. We have since learned that all planetary nebulae have them, in fact are in part defined by their presence, though not all are easily visible.

Like proper star names (Vega, Betelgeuse, and so on), proper names for nebulae, except for a famous few, are cumbersome. Instead, astronomers use catalogue numbers. The Dumbbell and Ring Nebulae are also respectively called Messier 27 and 57 (or M27 and M57) after Charles Messier, a French comet hunter who compiled a list of 103 nebulae and other fuzzy objects (most of which are star clusters and galaxies) in the 1780s. Many more planetary nebulae (and more than 7000 other objects) are known by their numbers in the "New General Catalogue" (the NGC) compiled in 1888 by J. L. E. Dreyer from hosts of discoveries, the Dumbbell, Ring, Saturn, and Helix also respectively called NGC 6853, 6720, 7009, and 7293. The NGC is actually an updated version of the great "General Catalogue" of non-stellar objects compiled by William's son John Herschel from his and his famous father's discoveries.

While the appearances of planetary nebulae share some common ground, notably a central star, there are great differences in detail among them, making them entertaining to see as the telescope is moved from one to another. The Dumbbell consists of two distinct lobes that define a "bipolar" structure. The Ring speaks for itself, though it is not round but quite elliptical, nor (like most planetaries) is it uniform, but is much brighter along its minor axis than along its major axis. The Helix, though a ring structure, consists more of two overlapping rings that make it look something like a crude bedspring. Other planetaries are quite round and yet others are filled, smooth disks. Some (actually, like the Dumbbell) have a "point symmetry," wherein one side is mirror-reflected into the other. Still others appear as nested rings, and many have extensive outer structures several times larger than the inner nebulae.

Finer details abound. The Saturn Nebula derives its name from the twin "ansae" – "handles" – that seem to project from the major axis of a complex elliptical structure. The interior of the Helix is filled with hundreds of comet-shaped structures that point away from the central star and that blend into the Helix Nebula's ring structure. Theories now being developed must unify this great array of structures and all the inner curious components.

The dimensions of the planetaries differ as much as their shapes. There are planetaries angularly so large it is hard to see them, while some are so small (or distant) that they look stellar. Physical dimensions require distances, which are difficult to determine. With some few exceptions, planetary nebulae are too far away, or their

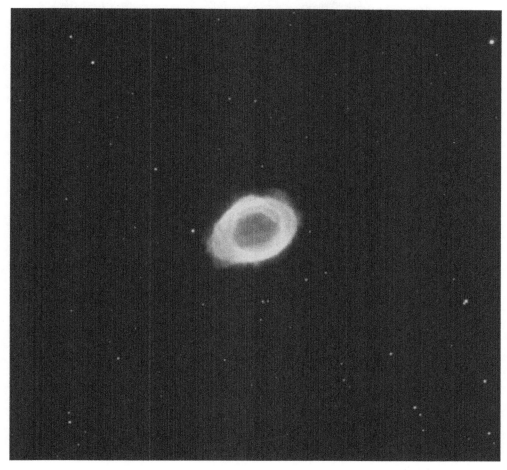

Figure 4.3. The Ring Nebula in Lyra, M57, has a different shape than the Dumbbell or the Saturn Nebula. Like that of the Dumbbell, its central star has a temperature over 100,000 K. [G. H. Jacoby, AURA/NOAO/NSF.]

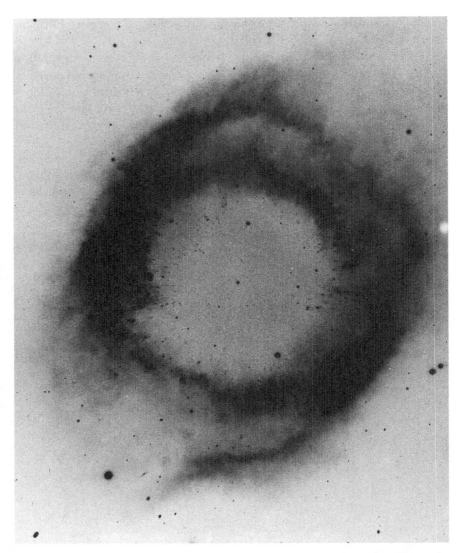

Figure 4.4. At a distance of only 300 light-years, and presented as a negative to enhance detail, the Helix Nebula, NGC 7293, in Aquarius, is the closest known planetary nebula. It is also one of the physically largest, stretching over 2 light-years across. Elongated knots of material in the central hole stretch away from the central star. [W. Baade, Palomar Observatory.]

stars too faint, for parallax measurement, and we must rely on a variety of indirect means. Still, we can tell that they range in diameter from several hundred astronomical units to monsters some light-years across that can span distances that typically separate stars.

The pairing between nebula and central star long ago demonstrated an intimate bond between the two. The true relation, however, is the stuff of our own century. We

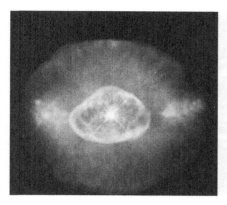 

Figure 4.5. The planetary nebula NGC 6826 (*left*) exhibits a double-shell structure as well as ansae similar to those seen in NGC 7009. This whole structure is surrounded by yet another shell (*right*) over four times the size of the image on the left, the huge halo the remnant of an earlier phase of the AGB star's wind. [*Left*: B. Balick (University of Washington) *et al.*, STScI, and NASA; *right*: G. H. Jacoby, AURA/NOAO/NSF.]

now know that the central stars are in the last stages of dying, and in doing so have marked their positions for us with wreaths made from the winds of the sky-stalking giants that the stars once were. It is the stars that make the nebulae glow. To do so they must be exceedingly hot, and in fact in general are hotter than any main sequence star. Welcome now to the realm of the hottest stars.

## Spectra

About 1850, the German physicist Gustav Kirchhoff described how a heated low-density gas seen by itself radiates emission lines, bright radiation at very specific wavelengths. For absorption lines to appear, the low-density gas must be in front of an incandescent source that produces a continuous spec-

Figure 4.6. (See also Plate VI.) The Helix Nebula's closeness together with the Hubble Space Telescope allows us to see the details of the extraordinary knots of matter in the central hole that blend to form the inner ring. [C. R. O'Dell and K. P. Handron (Rice University), STScI, and NASA.]

trum, a smooth spectrum with no breaks or gaps. The low-density gas absorbs radiation at specific wavelengths by raising the electrons to outer orbits from which they fall to re-radiate the energy as emission lines. The presence of emission lines therefore calls out "low-density gas."

When in 1864 the English astronomer William Huggins turned his spectroscope

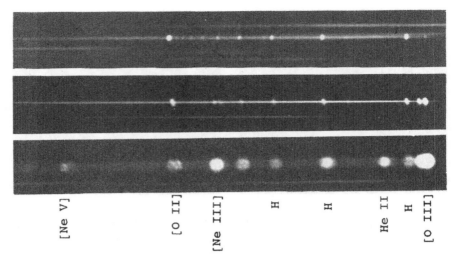

Figure 4.7. In these historic spectrograms taken by L. H. Aller in 1944 and 1945, everything in the telescope's field of view is spread into a spectrum (the stars recognized by their absorption lines). Since the nebulae radiate at specific wavelengths, each emission line makes an image of the entire nebula (note the structures). Huggins's three lines (which include one of hydrogen and two of doubly-ionized oxygen) are at far right. Forbidden lines are denoted by brackets; the ionization states are given by Roman numerals, beginning with the neutral state. The doubly-ionized oxygen lines are designated [O III]. The spectra are diverse. The [O III] lines are barely visible in Cn3-1 (*top*), but strong in IC 2149 (*middle*) and J 900 (*bottom*). The ionized helium (He II) line is very strong in J 900, but absent in the other two. [L. H. Aller, Lick Observatory.]

onto what is now called the "Cat's Eye Nebula," the planetary NGC 6543 in Draco, he saw three emission lines and realized immediately that Herschel's planetaries were made of heated gas. The shortest-wavelength line, which fell at 4861 Å in the blue-green part of the spectrum, belongs to hydrogen. At least in part, the planetary nebulae had to be made not just of gas, but of *hydrogen* gas. Another, stronger, hydrogen line at 6563 Å in the red part of the spectrum helps give many planetary nebulae a reddish color. The planetary nebulae are in fact archetypical astronomical emission-line objects, radiating a rich spectrum of bright lines at all the hydrogen and helium wavelengths as well as at a great variety of others that belong to different chemical elements and their ions.

These emissions provide firm evidence that the nebulae have been ejected by the star. If we look at the spread-out spectrum of a narrow slice of a nebula, the emission lines commonly appear split in two much like the hydroxyl lines in the clouds of molecular gas that surround the OH/IR stars. One component of the line is shifted to the violet part of the spectrum indicating that the gas is moving toward us, while the other is shifted to the red side, from gas going away from us (relative to the star, which has its own motion along the line of sight). The cloud must either be expanding or contracting. Contraction of something this large, with such a low gravity, makes no sense; expansion is the only viable theory.

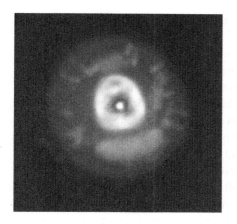

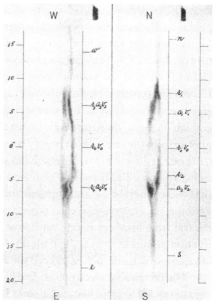

Figure 4.8. (*right*) This historic drawing of two views of the doubly-ionized oxygen lines was made by J. H. Moore in 1916 by placing a narrow slit across the face of the Eskimo Nebula, NGC 2392 (*above*; see also Plate VII). We then see a spectrum of a slice through the object. Each view of the emission line is split by the Doppler effect as a result of the expansion of the gas cloud. The right-hand spectral component of each comes from the receding far side of the object, that on the left from the approaching near side. The expansion speed derived is rather high for a planetary nebula, about 50 km/s. Note considerable other structure in the emission line that is caused by internal motions. [*Above*: AURA/NOAO/NSF, B. Balick; *right*: C. E. Moore, Lick Observatory.]

The expansion speeds are modest, typically about 20 km/s (though velocities up to 100 km/s are seen), and only a bit larger than those seen in the OH/IR stars. It is evident that the nebula has been ejected by the star, and from the expansion speeds and nebular dimensions, not all that long ago, only a few thousand years. Moreover, the nebulae cannot last terribly long, as the gas from a small young nebula should completely dissipate in only a few tens of thousands of years. The correct connection with the evolutionary predecessors was made once the Mira variables – the AGB stars – began to be understood, once we learned about their fierce winds. It was reasonable to assume that the planetary nebulae are the remnants of the winds and that the central stars are the AGB stars' now-exposed cores. All additional lines of evidence confirm the conclusion. Herschel's planetary nebulae are an intimate part of the flow of stellar evolution, of the progress of a star from birth to death.

But how do the central stars light the nebulae? The mechanism, not quite the same as envisioned by Kirchhoff, was worked out in 1928 by the grandmaster of the subject, Hermann Zanstra of Holland. To excite the hydrogen emission lines, the stars must be so hot that they radiate energy in the far, or high–energy, part of the ultraviolet spectrum. Given enough energy, an electron bound to an atom can be ripped away, the atom becoming ionized. An electron in the bottom rung of hydrogen's energy ladder requires the absorption of a photon with a wavelength shorter than 912 ångstroms, a dividing line called the "Lyman limit," radiation that would give a deadly burn to an exposed human.

As you heat a star, more energy pours out at shorter wavelengths. At a temperature of about 25,000 kelvin (roughly corresponding to the dividing line between the O and B stars of the main sequence), photons with wavelengths shortward of the Lyman limit emerge, photons that are capable of ionizing hydrogen atoms in the planetary nebula's gas out to the point at which all the photons are absorbed. If there are enough of them, they may even ionize all of the surrounding gas. Every ultraviolet photon from the star with an energy beyond that of the Lyman limit is capable of ionizing one hydrogen atom; the result is a sea of free particles, of protons and electrons, called a plasma. But because of the electrical attractions between the two kinds of particle there must also be a strong urge for *recombination*, that is, when an electron and a proton pass too close to each other, the proton captures the electron to produce a neutral atom once again, one capable of absorbing another central star photon. The fraction of ions in the gas depends on the competing rates of ionization and recombination, which are such as to keep the hydrogen in a very highly ionized state.

The recaptured electron can find itself on any of the possible orbits, or rungs of the energy ladder, and because it seeks its lowest energy, it jumps inward toward the nucleus, to a lower-energy rung, releasing its energy as a photon. The rungs are fixed in energy, so the energy, hence wavelength, of the emitted photon depends on the difference in energy between the two relevant orbits. As a result we see emission lines at specific energies or wavelengths, hydrogen producing emission lines at the same wavelengths that it absorbs in stars. Helium behaves the same way to produce other emission lines at different wavelengths, as do many of the lighter elements like oxygen, carbon, neon, and nitrogen.

Yet though we now know that hydrogen dominates stars, its emissions are not generally the strongest to be seen in nebulae. The other two emission lines seen by Huggins are usually both stronger and fall in the green part of the spectrum at wavelengths of 5007 Å and 4959 Å. They can be so powerful that many planetaries take on a distinctly greenish cast. Their chemical identification remained a mystery for 65 years after Huggins's discovery. For a time they were thought to arise from a heretofore undiscovered element, which was appropriately called "nebulium;" after all, helium had first been identified in solar spectra. Equally mystifying were the increasingly large numbers of unidentified lines that were found throughout the optical and the accessible ultraviolet spectrum (that part that the Earth's atmosphere lets pass through).

Ira S. Bowen of Mt Wilson Observatory cracked the problem in 1928 when he discovered that the original nebulium lines were produced by electron jumps between low-energy orbits of doubly-ionized oxygen. The nebular lines themselves were not observed in his laboratory; instead, the discovery was made by producing a complete map of the orbital structure of the ion by means of electron jumps that *were* visible. The relevant electron jumps that make the nebulium lines are very unlikely ones that the electrons are reluctant to make. The jumping-time for an excited hydrogen electron is a tiny hundred-millionth of a second. The electrons that make the nebulium lines, however, take a leisurely minute or more, and the emissions are therefore (erroneously) called "forbidden lines." Though not strictly forbidden to occur, the lines are seen only under special conditions within a low-density gas of great mass.

For an atom to produce emission lines, its electrons must absorb energy. Absorption of photons, as expressed by Kirchhoff, is one possibility. Mira variables have their hydrogen emission lines excited by shock waves generated by the pulsations. Bowen's forbidden lines are produced by collisions between ions and the energetic electrons that are freed from hydrogen in the initial ionization process. If the colliding electron has enough energy, it can knock the outer electron of doubly-ionized oxygen (the "active" one) into the appropriate higher orbit from which it then jumps back down to lower ones. Even if the downward process takes a long time, the electron has little else to do but wait it out, and eventually, down it goes. Once Bowen knew the mechanism, he could identify a great number of other forbidden lines that arise from oxygen, nitrogen, neon, argon, iron, and other elements. Some of the transition times are measured in hours.

Atoms are found in all manner of ionization stages, depending on the individual nebulae. Some planetaries have emission only from neutrals and lowly-ionized atoms such as singly-ionized oxygen and nitrogen. Others exhibit strong lines of doubly-ionized helium. In a few instances we even see multiply-ionized metals such as iron with six electrons removed. Considerable ultraviolet energy is needed for such ionization, indicating that the central stars can be very hot indeed. Moreover, the stellar temperatures must have a great range to produce an equivalent range in ionization level. Yet more, the ionization correlates with nebular properties. The smallest nebulae tend to have very low ionization levels, which grow with nebular diameter. The largest nebulae reverse the trend, implying that the stars first heat with time and then cool.

In spite of their seemingly odd natures, the spectra of forbidden lines are quite easy to interpret in terms of nebular conditions, allowing us to find the temperatures and densities of individual nebulae required to reproduce the strengths or brightnesses of the emissions. We typically find gas temperatures that are in the neighborhood of 10,000 K and densities that range from tens of thousands of atoms per cubic centimeter for small objects down to only a hundred or so for really big ones. Such densities are better than the best vacuums that can be produced on Earth. Like the Sun's low-density corona, the nebular gases do not follow the Stefan–Boltzmann law (which says luminosity per square centimeter of surface is proportional to the fourth power of temperature), so they actually radiate little energy. The temperature refers only to the speeds of the atoms and electrons in the gas, and so is properly called an "electron temperature."

Once we have the electron temperatures and densities of the nebulae, we can combine the theory for the production of the hydrogen and helium (and other) recombination lines with that of the forbidden lines and find chemical compositions,

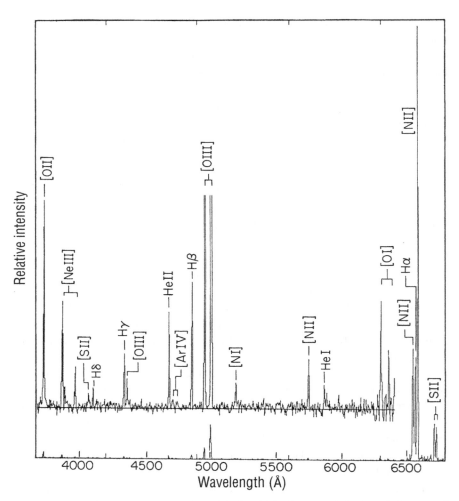

Figure 4.9. Modern spectra are presented graphically, with brightness (relative intensity) measured upward on the vertical axis. A planetary nebula called BV-1 has a complex spectrum rich in forbidden lines (denoted again by square brackets, with ionization states designated by Roman numerals). BV-1 also has 50% more helium than the Sun and 10 times as much nitrogen, which was made in the predecessor giant star before the nebula was born. The enormous nitrogen content is reflected in the great strength of the forbidden nitrogen lines near the red hydrogen line (Hα) at 6563 Å. [AURA/NOAO/NSF, author.]

the ratios of the number of the different kinds of atoms to the number of hydrogen atoms. Most planetary nebulae have compositions rather similar to that of the Sun. However, quite a few are highly enriched in helium (up to double solar), in nitrogen (sometimes by over a factor of ten), and in carbon. Nothing could be making elements within the relatively cool gas of the nebula itself; we are seeing the results of nuclear burning within a star. The splitting of the emission lines tells that the nebula must have come from a star and we know that a selection of AGB stars, including

Miras, are carbon-rich. The connection between the planetaries and the AGB stars is thus enormously strengthened. The carbon-rich nebulae must have been ejected by carbon stars, the carbon made by helium burning and brought up from below by convection before the ejection of the nebula. The excess helium in some nebulae must be related to hydrogen fusion in the AGB star's outer nuclear-burning shell.

The excess nitrogen is the result of nuclear burning by the "carbon cycle." The Sun and its kind run on the proton–proton chain. Above about two solar masses, however, near the transition of spectral class F into class A, carbon gets into the act and begins to create not just helium but other atoms as well. Deep in the stellar core, a $^{12}$C nucleus picks up a proton, making an isotope of nitrogen, $^{13}$N. This version of nitrogen, however, is radioactive, and quickly decays when one of its neutrons spits out a positron and becomes a proton, the nucleus thus returning to carbon, but this time $^{13}$C, which is stable. (The positron annihilates with an electron, producing two gamma rays). But the $^{13}$C can capture a proton too, making normal nitrogen, $^{14}$N, which picks up another proton to make radioactive oxygen, $^{15}$O, which in turn decays into stable $^{15}$N (and a positron). Forging ahead, the $^{15}$N grabs yet a fourth proton. Instead of making normal oxygen, $^{16}$O, the nucleus falls apart to carbon. But carbon has already picked up four hydrogen atoms, which are now glued together into a helium nucleus, which is ejected. As in the proton–proton chain, four hydrogen atoms have become one helium atom, here the carbon remaining unchanged. A secondary cycle can make more nitrogen and some actual oxygen.

The process must reach some kind of equilibrium, and therefore a steady level of freshly made nitrogen will exist. The carbon cycle in higher-mass stars completely dominates the proton–proton chain. The same process acts to produce energy in the hydrogen-burning shells of AGB stars. This is the nitrogen that makes it into the enriched nebulae, again brought up from below by convection before the creation of the nebula. Through these nebulae – and the extended shells around them – pass most of the carbon and nitrogen that reside in the Universe.

The nebular compositions let us look inside the giant-star predecessors of the planetaries and test theories of stellar structure, evolution, and all the processes that contribute to the ultimate abundances of the nebulae. So far, the observations crudely match the theories, which allow us to see, for example, that nitrogen-rich nebulae must arise from stars that initially had main sequence masses greater than about three times that of the Sun, which from the relation between mass and luminosity corresponds to stars near the transition where spectral class A becomes class B. Several things are still mystifying, however. For example, the stars can cycle upward much more helium than can be expected. We still do not understand AGB stars all that well.

## The central stars

From the surface of the Earth we observe only that portion of the electromagnetic spectrum passed by our protective atmosphere, which blocks most ultraviolet and infrared. These spectral regions, however, carry an enormous amount of information not available optically, and yield data on both high- and low-energy processes. The central stars of planetary nebulae are so hot that they radiate most of their energy in that invisible ultraviolet. Vast sums have been spent to send complex telescopes into space to view portions of these hidden regions, often with spectacular results. The nebulae provide a simpler way, one that was used long before the advent of spaceflight to learn the natures of the central stars.

The temperature of a star can be found from the wavelength at which its background continuous spectrum is the brightest (from the Wien law). But the central stars of planetary nebulae are all so hot that peak emission falls so far into the ultraviolet that it cannot be observed. In the optical part of the spectrum the background continua of all these stars look very much alike and temperature cannot be discriminated; for that matter, even in that part of the ultraviolet accessible from space, the differences are not all that great. We must look to very short waves by a method that was pioneered by Zanstra.

Each ultraviolet photon more energetic than the Lyman limit will ionize one hydrogen atom; each hydrogen atom must subsequently recombine with a free electron; and each recapture will quickly lead to an electron jump that terminates on the lowest-energy orbit. However almost all the hydrogen atoms have their electrons in that lowest orbit; consequently any photon created by a jump to the lowest orbit will almost immediately be reabsorbed, kicking the electron back upstairs. The photon finds itself trapped, bouncing from one atom to another. The only way out is to jump first to a higher-energy orbit in which there are too few electrons to reabsorb the new photon, which then escapes the nebula. By this process, all recaptures ultimately wind up in the second orbit before jumping to the bottom. But jumps to the second orbit produce the optical emission lines. All we need therefore do to count the number of ionizing photons leaving the star per second, those shortward of 912 Å, is to count the number of optical hydrogen photons leaving the nebula per second.

The number of ultraviolet photons leaving the star per second by itself gives us little information, as it depends not only on the star's temperature, but on its radius as well, which is not known. However, we can also make a measure of the star's brightness in the optical spectrum, where its apparent visual magnitude gives the number of yellow photons leaving the star per second, which depends on temperature, stellar radius, and distance. But both measures, those in the ultraviolet and in the yellow, depend on radius and distance in the same way. So if we just divide the number of ultraviolet photons leaving the star per second by the similar number of yellow photons, radius and distance drop out and we have a measure of temperature alone! Zanstra thereby was the first to determine the temperatures of the central stars of planetary nebulae; in his honor, values so found are called "Zanstra temperatures." The technique can be extended to helium. Helium is much more difficult to ionize than hydrogen. It takes photons with wavelengths shorter than 512 Å to strip away one electron, and shorter than 228 Å to get rid of two. Consequently, the optical ionized helium lines count photons in the extreme ultraviolet, and we can use the same trick to get temperature.

The problems with the method are that some ultraviolet photons might escape the nebula and therefore not be counted and that the spectrum of the starlight in the ultraviolet might not be as simple as expected and may not follow the rules on which the Stefan–Boltzmann and Wien radiation laws are based. It is only in a very few nebulae that the very short-wave ionizing photons escape, however, and the helium temperatures should be close to the truth, even if the hydrogen temperatures are not.

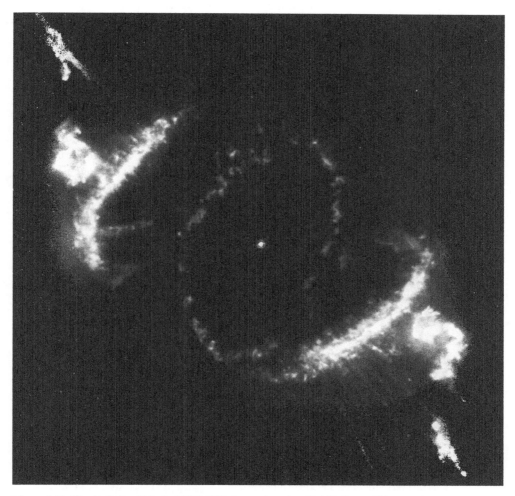

Figure 4.10. The Cat's Eye Nebula, NGC 6543, is seen here in a Hubble Space Telescope image taken in light radiated by ionized nitrogen. This enormously complex object has ansae reminiscent of those of NGC 7009 that seem to spray from the central star, which has a relatively modest Zanstra temperature of about 47,000 K. [J. P. Harrington and K. J. Borkowski (University of Maryland), STScI, and NASA.]

Zanstra temperatures begin at a low of about 25,000 kelvin, just that expected for the minimal ionization of a planetary nebula, giving credence to the method. As high as this value is, it is still comparable only to that of the cooler O stars. But now climb the temperature ladder. The central star of William Huggins's favorite nebula, NGC 6543, lies at 47,000 kelvin, near the limit of the hottest main sequence star, and the central star of Herschel's original nebula NGC 7009 is at a much higher 80,000 kelvin. But we have not even begun to reach the top. The Ring and Helix Nebulae both have central stars that are around 130,000 kelvin.

The increasing temperatures have an odd effect. As temperature climbs, the stars become less visible, more and more of their radiation coming out in the ultra-violet where it serves to brighten the nebula. As a result, the contrast between nebula and star goes down, eventually hitting the point at which the star nearly disappears. For decades some planetaries have had the dubious distinction of having no visible central stars at all, one that has only recently fallen as a result of new observational techniques that allow the computer-subtraction of the nebula from an image, and from imaging in the ultraviolet (where the star is much brighter) with the Hubble Space Telescope. The best-known "star" in this category is that of NGC 7027 in Cygnus, a planetary with no special name. It is, however, one of the best known of all to professional astronomers, as it is one of the brightest in the sky and is ionized to a very high level, implying an extremely hot star, one now measured at a whopping 175,000 K. NGC 2440, a classic bipolar nebula in the constellation Puppis, whose central star was also only recently revealed, comes in at the highest confirmed temperature of 220,000 K. Less secure values go to nearly 300,000 kelvin, and there are probably a few unanalyzed ones that go even higher.

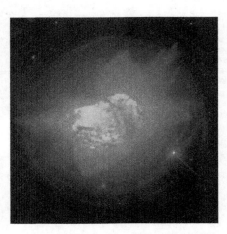

Figure 4.11. (See also Plate VIII.) NGC 7027 is a complex, dusty nebula that has one of the highest central star temperatures known, so high that very little of the star's light is radiated in the optical. The nebula is rich in carbon, implying that its progenitor was an AGB carbon star. The concentric shells represent earlier episodes of mass loss. [H. Bond, STScI, and NASA.]

It is easy to be awed by a beautiful planetary nebula as seen through the telescope. The central stars, however, in spite of the accolades given above, still seem distinctly unimpressive. But that is a result of most of their energy exiting in an invisible part of the spectrum. When we calculate the actual total luminosities of these stars (the amounts of energy released per second) from the measure of the number and energies of the ultraviolet photons leaving the star per second, these stars are among the brightest in the Galaxy, with luminosities that can well exceed 10,000 times

that of the Sun. Any planets that survive the giant-branch phase of stellar evolution would be thoroughly fried by such brilliant central suns. These seemingly modest stars reveal both the limitations of our human senses and the need for astronomers to have access to other wavelength regions. Equally impressive is the range of stellar luminosities, which extend downward to only a few times solar for the dimmest of them.

## Spectra and the HR diagram

Different kinds of stars are generally compared with one another through the HR diagram, on which we usually plots stars according to their absolute visual magnitudes and their spectral classes. The central stars of planetary nebulae, however, do not fit, as their spectra are very different from anything we have seen so far, and the temperatures run much higher than those of any main sequence star. As a result, the diagram must be modified, and the stars plotted by their temperatures (which up to a point correspond to spectral classes) and actual luminosities.

When so plotted, most of the central stars fall within a broad, fairly well-defined path that extends to the left of the most luminous AGB stars toward very high temperatures at more or less constant luminosity (the "horizontal portion") and then reverses, falling downward and to the right (the "descending portion"), the stars here becoming dimmer and cooler. On the average, the planetary nebulae associated with the stars get progressively larger from right to left on the horizontal portion, then from left to right on the descending portion; the tiny ones are closest to the AGB, while the huge ones like the Helix are down toward the bottom. Since we know that the nebulae are expanding, the nebulae – and their stars – must therefore be youngest (relative to the birth of the nebulae) at the right-hand end of the horizontal portion and oldest at the right-hand edge of the descending portion. The path within which we find the central stars must therefore define a broad evolutionary track that shows how the stars change their characteristics with time.

The upper right end of the path lies at the same luminosity as the observed top of the AGB. Though a wide observational gulf separates the AGB from the planetary central stars (one that must be bridged in a semi-hidden state), it is evident that the central stars must be the descendants of these advanced-evolution giant stars, especially since we know that the AGB stars are furiously losing mass. The central stars are the old, now-tired, nuclear-burning cores of the giants. The coolest stars – the AGB giants – perform the remarkable feat of transforming themselves within a short interval into the hottest stars.

Calculation of stellar diameters from temperature and luminosity shows that the stars also become smaller as they proceed from right to left (from low temperature to high) along the horizontal portion and then downward along the descending portion. At the end, they are comparable to the white dwarfs in size. The giants transform themselves not only from the coolest to the hottest stars, but from large to small as well.

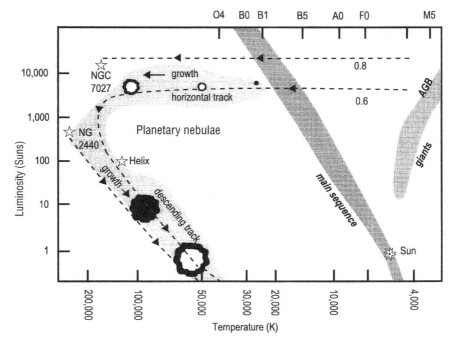

Figure 4.12. The central stars of planetary nebulae occupy a bow-shaped zone in the extended HR diagram (with luminosity plotted against temperature). Within it, the nebulae become larger from upper right to the bend in the bow (the horizontal track) and then from the bend to lower right (the descending track), the increasing radii defining a flow of evolution. Three of the hottest central stars are plotted within the band. Theoretical evolutionary paths (dashed lines) for spent cores of 0.6 and 0.8 solar masses fit well with the observed band. Higher-mass stars have higher luminosities on the horizontal band, lower luminosities on the descending portion.

Because of the huge range of temperature and luminosity, we find an enormous range of spectral characteristics: in the domain of the hottest stars, we deal not with just one kind of star but with many. While many of the stars have absorption lines typical of more ordinary high-temperature main sequence stars, the spectra of the brightest planetary central stars, those on the horizontal portion of the evolutionary pathway, also display emission lines of carbon, helium, and other elements that indicate outflowing clouds of circumstellar matter.

Especially in the ultraviolet, central stars can have peculiar composite spectrum lines that consist of both absorption and emission components. The emissions are at the correct wavelengths (taking into account the overall radial velocities of the stars), but the absorption features are considerably shifted toward shorter wavelengths. Such lines were first observed in the high-mass star P Cygni, and have been called "P Cygni lines" ever since. The absorptions must be coming from gas moving at high speed directly toward the observer in front of the stars in order to absorb starlight, and clearly demonstrate the presence of strong outflowing winds. The rest

of the gas, that not in front of the stars and flowing in all directions, produces the emissions. Since the gas making the absorptions is coming right at us, the absorptions' maximum Doppler shifts give the winds' velocities, and from the amounts of radiation the absorptions extract from the spectra, we find the stars' rates of mass loss.

In planetary youth, at the lowest stellar temperatures, sometimes about all we can see in optical spectra are overlapping lines, the continuous spectrum practically buried beneath them, and no absorptions at all. Though the chemical analysis of a spectrum this complicated is difficult, we have learned that the hydrogen abundance can be very low (if there is any at all) and that the carbon abundance can be relatively high, with a ratio of carbon to helium 100 or more times greater than solar. Such stars have been nearly stripped of their outer hydrogen envelopes, and the carbon must have come from the earlier burning of helium in the preceding giant stars. Wind speeds are a few hundred kilometers per second, which, though much higher

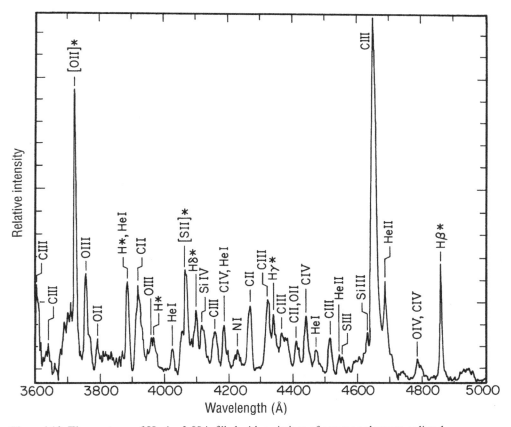

Figure 4.13. The spectrum of Henize 2-99 is filled with emissions of common elements radiated by a strong, encompassing wind that quite literally hides the star. The Roman numerals give ionization stages. Lines with asterisks belong to the nebula. [AURA/NOAO/NSF, author.]

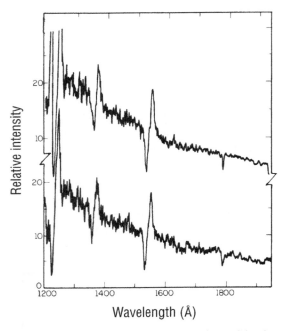

Relative intensity

Wavelength (Å)

Figure 4.14. Ultraviolet spectra of the central star of the planetary Abell 78, taken with the *International Ultraviolet Explorer* satellite, display "P Cygni lines" of carbon (at 1550 Å) and oxygen (1371 Å), in which bright emission lines are flanked to the short-wave side by absorption lines that indicate absorbing matter flowing directly at the observer. [*IUE* spectrum by J. B. Kaler, W. A. Feibelman, and NASA.]

than observed in Mira variables, are modest for planetary nebula central stars. Mass loss rates, however, typically about a hundred millionth of a solar mass per year, are nowhere near as great as in Miras, though they are still some 100,000 times larger than that emerging from our relatively placid Sun.

As we look toward higher temperatures the emission spectra become simpler, though many are still dominated by helium and carbon, and the level of ionization changes in response to increased temperature. Some hot stars have practically no wind features, but others, at the high-temperature end of the scale, show powerful lines of highly ionized oxygen in addition to carbon features. Wind speeds pick up to several thousand km/s toward the high-temperature end of the evolutionary path, far greater than those of the Mira variables.

As the stars turn the corner in their evolutionary pathway and – aging – begin to descend to lower temperatures and luminosities, the winds die down. The low-luminosity stars, those that are descending downward on their evolutionary paths toward the white dwarfs, display little more than absorption lines. From these, astronomers can determine accurate stellar temperatures, which are more or less in line with the Zanstra temperatures. The absorptions indicate some interesting behavior. A few, like the central star of the Helix nebula, are deficient in helium, while others may be anomalously rich in it, characteristics that will play a major role when we look later at the smallest stars.

Theory explains much of what we see. As a star nears the top of the AGB giant branch, its wind blows ever-more fiercely, becoming a "superwind" by which it strips itself down nearly to its nuclear-burning core, a ball of carbon and oxygen surrounded by nuclear-burning shells that, in turn, are surrounded by a low-mass skin of hydrogen. The shrinking core has to this point generated the ever-increasing stellar luminosity. But now the electrons – stripped from their atoms in the terrible heat – come into play.

Within a gas, at any given temperature, atomic particles move at a variety of speeds; the greater the temperature, the higher the average, and the higher the

maximum. In the world of the atom, however, "particles" take on wave-like natures. In part, because the electrons behave as much like little waves as they act like tiny balls, the electrons that move at a specific speed can get no closer, becoming "degenerate." They therefore produce an outward pressure that first slows, then stops, the core's contraction.

The star now stabilizes and can get no brighter. What is left of it then leaves the AGB, hidden within a shroud of dust that condenses out of the departing wind. The wind continues now at a lower rate, but begins to speed up, thinning the hydrogen envelope from above at the same time nuclear burning in the hydrogen shell thins it from below, and the star's surface heats, sending it to the left on the HR diagram at constant luminosity, the stellar radius shrinking commensurately with the higher temperature. At the same time, the lost mass is expanding, and when the star hits 25,000 K or so, the now-larger cloud becomes ionized and illuminated. As the star's surface temperature tops 100,000 kelvin, the nuclear burning that is still taking place in the thinning shell begins to shut down. The star then reverses itself on its evolutionary pathway and – now cooling and dimming – enters the descending path.

The broad band of the observational evolutionary path is produced mostly by a range of stellar masses. As the initial mass of the star (the mass with which the star was born on the main sequence) increases, so does the mass of the core, the part that eventually turns up at a planetary nebula's center. Theoretical calculations show that the masses of stars within the observational band go from about 0.55 solar masses to a bit over one solar mass. On the horizontal portion, the higher the mass of the star the greater its brightness. But in the descending portion, the relations reverse, the higher-mass stars the fainter because their greater gravity squeezes them down more and gives them lower surface areas.

By the time the stars have cooled to about 75,000 kelvin or so, the nebulae have expanded to the point where they are so thin that they can no longer be seen. The gas departs to blend with the matter of interstellar space, and the stars are left behind to cool into the realm of the white dwarfs.

## Sculpting a nebula

The picture painted above seems simple, but the planetaries themselves are not. The mechanism described – the mass lost by the AGB star illuminated by its exposed core – suggests that all planetary nebulae should be uniform expanding shells. Instead, the nebulae come in an almost bewildering variety of shapes and forms that need to be explained. More is going on than just the ionization of the departing wind.

The fast, hot wind that develops from a newly exposed core is the critical factor. Imagine the last portion of the dusty wind – that produced by the "superwind" of the giant star – slowly expanding away from its central source. At first the star in the center is effectively invisible, surrounded by a dusty obscuring shroud. During this

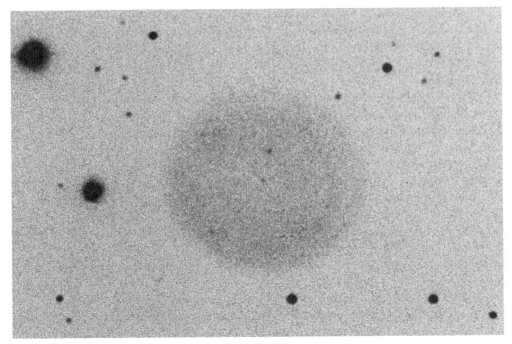

Figure 4.15. A large planetary nebula, Abell 16, is beginning to merge with the interstellar medium, the picture presented as a negative for improved contrast. Note the faintness of Abell 16's central star as it descends toward the realm of the white dwarfs. [G. H. Jacoby, AURA/ NOAO/NSF.]

latent-planetary period, the star's surface heats from 3000 kelvin or so across the HR diagram, passing through what would be spectral classes K, G, F, A, and B, all changes invisible to the eye. But the star, too cool to ionize its surroundings, can illuminate the expanding nebula's dust by reflection. Several of these "proto-planetary nebulae" have been found, the hidden star in the center soon to become visible as the surrounding shell expands. In the incredibly dusty "Egg Nebula" we see dozens of concentric shells of reflecting dust that are somehow related to mysterious episodic ejections that take place every two or three centuries.

At some point not yet known, the shrinking star begins to produce a much thinner, but faster, wind. This new wind slams into the earlier fleeing superwind gas and shovels it up, compressing the inner edge and forming a much thicker shell within the outer one. Now when the star hits its critical temperature of 25,000 K, it ionizes a dense ring set within the outer dusty gas. When the whole system expands enough for the photons to leak into the outer portion of the nebula, we see a thick inner shell encased in an outer one. If there are enough ionizing photons, and the inner stuff expands to sufficient thinness, we might even see the vast outer halo of matter that was lost by the AGB star before the superwind phase, its once-outer envelope. The concept neatly explains the so-called double shell nebulae and the huge halos seen around such notable nebulae as NGC 6543 and NGC 6826.

The shape of the planetary nebula will depend on the distribution of the super-wind matter lost by the AGB star and on the age of the object, the time since the nebula itself was born. If mass is lost uniformly (apparently a rare instance), we will see a near-perfect nebular sphere that appears projected on the sky as a bright ring. But we know that the lost mass cannot be completely uniform, if for no other reason than that AGB stars do not appear to be spherical. Assume that the giant's super-wind is somewhat thicker at the star's equator than it is at the poles. The fast wind from the exposed core will blow more easily through the poles than it will at the equator since it meets less resistance. The shoveled matter will therefore take on an elliptical shape, with the long axis aligned toward to the star's rotational axis. As time proceeds, the wind may even break out of the superwind mass and blow bubbles per-pendicular to the equator. If the superwind is very strongly enhanced at the equator, so much so that the shoveling effect is nearly halted there, we see a wasp-wasted bipolar nebula.

Less certain, in some objects this "snowplow effect" might be focused for a time to produce high-speed jets of gas that pour out the polar axis. These hit the sur-rounding gas and heat it, possibly producing the ansae, the blobs that appear like the jug-handles on the Saturn Nebula, NGC 7009. The comet-like knots in objects like the Helix could be made when the hot star ionizes the leading edges of dense con-densations left behind in the turbulent battle between the fast and slow winds.

This theory, while perhaps explain-ing the nebulae, just puts off the ulti-mate question of what produces the non-uniform distribution of the wind in the first place. The original star's rotation may play a role. The bipolar nebulae tend to stick to the plane of the Galaxy where the higher-mass B stars are created. There are almost none in the ancient galactic halo, where no new star formation goes on, and it is well known that the higher-mass stars are rapid rotators. A stellar companion may also be important, and if one is close enough, it could strongly influence the wind from the huge giant by raising tides that help remove mass and direct its flow. In most stars the orbital planes and rotational axes were probably long ago gravitationally forced to align

Figure 4.16. (See also Plate IX.) The Egg Nebula in Cygnus, observed by the Hubble Space Telescope, consists of dozens of expand-ing dusty shells. Light from a warm star hidden in a thick disk illuminates the surroundings through the disk's poles. Eventually, the star's wind will compress its surroundings to a shell that will be ionized as the star heats to a tem-perature beyond 25,000 K. [R. Sahai and J. Trauger (JPL), STScI, and NASA.]

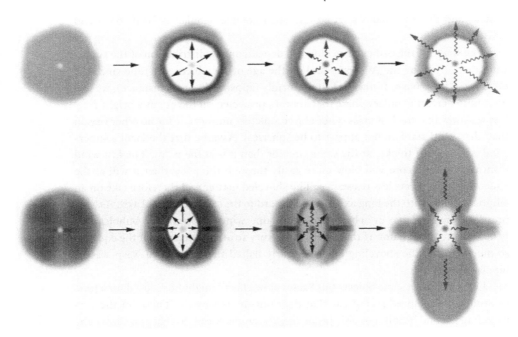

Figure 4.17. Differences in planetary nebulae might be caused by compressing winds that are focused by the way in which mass was originally lost. If mass is not lost in a preferential direction (*top row*), the fast compressing wind will produce a spherical planetary nebula. If it is lost in the thick disk, the fast wind will expand along the poles, resulting in a strongly bipolar nebula (*bottom row*). A variety of other possibilities depend on mass-loss distribution and viewing angle. [Adapted from *Cosmic Clouds* by J. B. Kaler © 1997 by Scientific American Library. Used with permission of W. H. Freeman and Company.]

themselves, so it would be natural for the thick AGB winds to align with the equatorial planes.

Even bodies as small as Jupiter-sized planets might stir the AGB giant enough to produce a non-uniform wind. Here would be the ultimate irony. An introduction to planetary nebulae is immediately followed by a disclaimer that they have nothing to do with planets. Perhaps, however, a star's planets might be crucial to forming the nebulae; moreover, the destruction of the inner planets by the evolving star might even add to the dust that hides its horizontal progress across the HR diagram.

The binary – double star – concept is supported by the "point symmetric" nebulae. The axes of rotating bodies commonly wobble, or "precess," like that of a spinning top. The Earth is a bit larger across its equator than through its poles. The Earth's axis therefore wobbles over a 26,000-year period as a result of the gravitational action of the Moon and Sun on the equatorial bulge. Orbits can precess too, the Moon's orbit wobbling with an 18.6-year period. If the orbit of a companion to a windy AGB star is not quite aligned with the star's rotation axis and it precesses, it may also wobble the blowing wind, making the wind flowing out of the poles flow first one way and then another. Planetary nebulae are memories of the past in that they illuminate the way in which giant stars whittle themselves down in the course of their evolution.

The stars on the descending track now seem destined to cool into the realm of the white dwarfs, and most go quietly. But the intimate relation between the coolest and the hottest stars is not entirely complete. Cool stars transform themselves into hot stars. Strangely, the reverse may also be possible. The members of a small subset of planetary nebulae contain inner, irregular shards of gas that are completely deficient in hydrogen and are made up mostly of helium. These central structures can only have come from a star that had first completely lost its exterior envelope. Yet the developing AGB star still has a hydrogen envelope while it is beginning to produce the planetary.

When the star was nearing the end of its term in the domain of the planetaries, at a time when it was fading, and the large nebula was dissipating into space, the helium shell that surrounded the spent carbon core apparently fired up and began to burn, as it periodically does within AGB stars. The event produced so much energy that it puffed the nebula's central star back to giant proportions, the hot star temporarily becoming cool once again. During this phase of low surface gravity, the star stripped itself of its remaining hydrogen skin – if it had any left at all by that time – and then began to pump out some of its helium underlayer. Thousands of years later, the star has re-evolved, the cool star again becoming hot, and we now see the remnant of the helium wind closely enveloping it.

These graceful objects that surround at least one set of the hottest stars (we will encounter another later, neutron stars that are even hotter) are probably not done with their surprises. They are telling us a great deal about their forebears, the coolest stars, and even more about their successors, one set of the smallest stars. Before these are addressed, however, we must look at the high-mass counterparts to the stars already examined, those at the upper end of the main sequence – the most luminous stars – and then at their evolved progeny, the largest stars, the supergiants. Only then will we return to the smallest stars to see what kinds there are and to look at the ultimate in stellar death.

Figure 4.18. HB-5 is a strongly bipolar nebula that seems to have been produced by a fast wind acting upon a severely asymmetric distribution of mass. It is a new planetary with a relatively cool central star. Note the rings that are reminiscent of the Egg Nebula. [B. Balick (University of Washington), V. Icke (Leiden University), G. Mellema (Stockholm University), STScI, and NASA.]

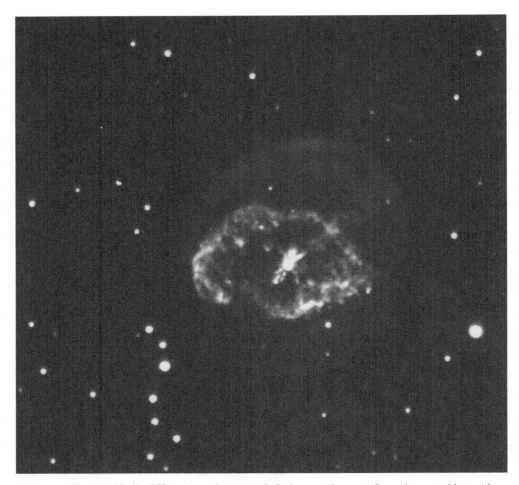

Figure 4.19. Abell 78 is a large planetary nebula that contains a set of complex central knots (those right next to the star) that have no hydrogen. They are made mostly of helium apparently ejected from the surface of the central star when it briefly expanded and returned to the giant branch of the HR diagram to produce a planetary within a planetary. [G. Jacoby, AURA/NOAO/NSF.]

# Chapter 5

# The brightest stars

The central stars of planetary nebulae are brilliant, as luminous as their predecessor giants at the top of the AGB, both kinds of stars reaching luminosities over 10,000 solar. In both instances the stars fool us, as they radiate light in portions of the spectrum that are invisible to us, the giants producing a great percentage of their energy in the infrared, the planetary central stars proportionately even more in the ultraviolet. If our eyes were sensitive to infrared, the coolest Miras would be some four magnitudes – a factor of 40 – brighter than we see them, and if we could observe all the ultraviolet (without the nebula absorbing any radiation), the hottest planetary nebula stars would be over 1000 times more luminous than they shine to our eyes.

But bright as they are, neither the AGB stars nor their hot successors win the luminosity prize. They are briefly bright because their advanced state of evolution allows the release of suspended gravitational energy and more advanced nuclear burning. To find the brightest stars we look not to evolution, but to mass. As stars descend downward from the Sun to the red dwarfs, they drop in mass to a limit of 0.08 that of the Sun and a low of about a millionth of a solar luminosity, the result of low interior temperatures produced by small compression. Ascending the main sequence from the Sun, the stars behave oppositely, becoming more massive, and as mass increases so does interior compression, temperature, and hence luminosity. At the top of the main sequence, among the O stars, where the stellar masses climb from 10 times solar to over 100, apparent visual magnitudes reach $-5$ to $-6$, well in excess of the visual luminosities of AGB or planetary stars. They are so massive, and so hot inside, that their cores run not on the proton–proton chain, as does the Sun, but, like all dwarfs from around class F0 and hotter, on the carbon cycle, in which carbon is used as a nuclear catalyst.

At the limit, at a spectral class of O3, a 100 or so solar mass main sequence star will have a surface temperature of nearly 50,000 kelvin, falling well within the range of the cooler planetary nebula central stars. As a result, these hot dwarfs will also radiate much of their energy in the ultraviolet part of the spectrum. Were that accounted for, the stars would be another four magnitudes brighter, the corrected (bolometric – see below) magnitudes reaching −10, a million times more luminous than the Sun. Evolution can make them even brighter, opening the door to a new realm.

## Who are they?

None of the faintest stars is even close to being visible to the naked eye, nor, because they radiate so much in the ultraviolet, are any of the hottest stars. Only a few of the coolest stars, the advanced giants like Mira and Chi Cygni, can easily be viewed, and these only because they are relatively close to us. We do not have such a problem with the "brightest stars," some of which can be seen even in modest lighting.

First distinguish between those stars that are bright to the eye because they are truly luminous and those that are bright because they are close. The seemingly brightest star in the sky, Sirius in Canis Major, is simply nearby, its parallax showing it to be only 9 light-years away. Though of class A and actually 25 times more luminous than the Sun, it is far from extraordinary. The same can be said for most other luminaries, such as Lyra's Vega, Aquila's Altair, and Alpha Centauri, whose brighter component (it is a double) is a truly modest solar-type star that is just next door.

Now, however, examine Cygnus, the celestial swan. His tail is marked by Deneb, a white class A first magnitude star that also marks the northeast apex of the Summer Triangle. Like Vega, it is bright to the eye but, unlike Vega, it is 2600 light-years away! Deneb has an absolute visual magnitude of −8.4, 200,000 times the visual luminosity of the Sun. If actually placed alongside Vega, Deneb would shine in our sky at apparent visual magnitude −9, fifty times brighter than Venus at her best and comparable to a crescent Moon. It would easily be visible in full daylight, and would cast clear shadows at night.

With Deneb the entry point, we leave the ordinary stars of the main sequence and even those of the giant branches and enter a strange realm of enormous and massive stars that shine with immense power. Fueled by huge nuclear engines, these spectacular stars quickly transform themselves from one kind to another as their nuclear furnaces greedily consume fuel. Beginning their lives on the main sequence, they become not giants but vastly larger *supergiants* that can take on an almost bewildering variety of forms. Deneb has left the main sequence and is the most luminous class A supergiant known in the Galaxy.

The winter sky of the northern hemisphere contains the supergiant that is brightest to the eye, Rigel, shining in Orion's foot at apparent magnitude zero. Though not as luminous as Deneb, its absolute visual magnitude of −6.7 (40,000 times brighter than the Sun) still ranks it among the top B stars. Six times more visually luminous is hotter Zeta-1 Scorpii, a B1 supergiant that does not help dominate its constellation only because it is eight times farther away than Rigel. Though at an immense distance of 6300 light-years, Zeta-1 Scorpii, at apparent visual magnitude 4.7, is still easily visible to the naked eye.

## The brightest stars

For each spectral class the table lists the brightest star that is known in the Galaxy, the most luminous with either Bayer names or Flamsteed numbers, and some of special interest. "LBV" means "luminous blue variable." All are supergiants except for Theta-1 Orionis-C, which is still on the main sequence. "Location" refers to the OB association or constellation of residence.

| Star | Apparent visual mag. ($V$) | Spectral class[a] | Absolute visual mag. ($M_V$) | Absolute bolometric mag. ($M_{Bol}$) | Location | Distance (ly) | Special properties |
|---|---|---|---|---|---|---|---|
| HD 93129A | 7.0 | O3 If | −7.0 | −12.0 | Carina | 11,200 | most luminous |
| Zeta Puppis | 2.3 | O4 Iaf | −5.9 | −10.2 | Puppis | 1400 | runaway star |
| Theta-1 Orionis-C | 5.1 | O6 V | −5.1 | −8.9 | Orion OB1 | 1600 | in Orion Nebula |
| Tau Canis Majoris | 4.4 | O9 Ib | −7.0 | −10.1 | NGC 2362 | 4900 | binary |
| Cygnus OB2 #12 | 11.5 | B5 Ia⁺e | −10 | −10.9 | Cyg OB2 | 5700 | 10 visual magnitude extinction |
| Eta Carinae | 6.2 | B0 0 | −10 | −11.9 | Carina | 8200 | LBV |
| P Cygni | 4.8 | B2 Ia–0 | −8.6 | −9.9 | Cyg OB1 | 7000 | LBV, loses $4 \times 10^{-4}$ solar masses per year |
| Zeta-1 Scorpii | 4.7 | B1.5 Ia⁺ | −8.7 | −10.8 | Sco OB1 | 6300 | loses $5 \times 10^{-5}$ solar masses per year |
| Rigel | 0.12 | B8 Ia | −6.7 | −7.3 | Ori OB1 | 775 | |
| S Doradus | 8.6 | A5 0 | −9.8 | −9.8 | LMC | 170,000 | LBV |
| Deneb | 1.25 | A2 Iae | −8.4 | −8.6 | Cyg OB7 | 2600 | loses $3 \times 10^{-10}$ solar masses per year |
| 6 Cassiopiae | 5.43 | A3 Ia⁺e | −8.3 | −8.4 | Cas OB5 | 8200 | shell star |
| IRC+10420 | ... | F8–G0 Ia | ... | −9.2 to −10.2 | Aquila | 15,000? | dust-enshrouded IR star |
| AG Carinae | 6 | O 0–F 0 | ... | −10.7 | Carina | 20,000 | LBV |
| Rho Cassiopiae | 4.54 | F8 Ia | −9.6 | −9.6 | Cas OB5 | 8000 | odd variable |
| HR8752 | 5.10 | G0–G5 Ia–0 | −9.3 | −9.5 | Cep OB1 | 11,000? | variable, 0.4 apparent visual mag; shell |
| RW Cephei | 6.65 | K0 Ia–0 | −9.4 | −9.6 | Cep OB1 | 11,500? | |
| Mu Cephei | 4.08 | M2 Iae | −7.3 | −8.5 | Cep OB2 | 2000 | SRc–Lc |
| VV Cephei | 4.91 | M2 Iae | −8.0 | −9.5 | Cep OB2 | 2000 | Lc, eclipsing |

*Notes:*

[a] Roman numeral I denotes that the star is a supergiant; Ia that it is a bright supergiant; Ib a less-bright supergiant. Class 0 (zero) denotes that the star is an even brighter hypergiant.

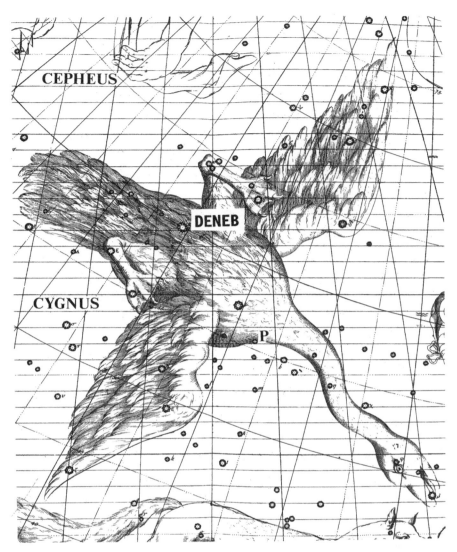

Figure 5.1. Deneb, the first magnitude star at the tail of the Swan, passes overhead in northern summer evenings. An A-type supergiant, it is one of the most luminous stars in the Galaxy. P Cygni, a brilliant B star whose temperature-corrected ("bolometric") luminosity is 1.3 magnitudes brighter than Deneb and is furiously losing mass, is marked with a P. [From Flamsteed's *Atlas Coelestis*, 1781 edition, courtesy of the Rare Book and Special Collections Library, University of Illinois at Urbana–Champaign.]

Back now to the main sequence and to obscure "Cygnus OB2 #12," visually the most luminous star known in the Galaxy, glowing at absolute visual magnitude $-10$. Though no farther than Zeta-1 Scorpii, it is nowhere close to being visible to the naked eye, appearing at apparent visual magnitude 11.5 as a result of being obscured by thick clouds of dust that lie along the line of sight. If there were no dust, Cygnus

OB2 #12 could be seen with the naked eye even if it were 60,000 light-years away. If at the distance of the nearest star, Alpha Centauri, a mere 4 light-years away, this single star would be some ten times brighter in our sky than the full Moon, and one could read by its light. Eta Carinae, in the southern hemisphere, is comparable – though, as we will see, vastly stranger.

But once more enter the invisible radiation, this time formalizing it. To indicate the total radiative output of a star, astronomers apply a "bolometric" (from Greek, meaning "measure of rays") correction to the visual magnitude, the result called the "bolometric magnitude." The coolest stars radiate almost all their energy in the invisible infrared, that is, their heating ability far outshines their visibility, the bolometric magnitude over four times brighter than the visual. For temperatures higher than solar, the hotter the star, the more invisible ultraviolet is produced. Rigel's correction is over half a magnitude, making the star 66,000 times more luminous in total power than the Sun. With a bolometric magnitude of nearly −11, hotter Zeta-1 Scorpii is 25 times brighter and, at bolometric −11.9, Eta Carinae becomes one of the Galaxy's most brilliant stars, 4.5 million times more radiant than the Sun. The hotter O stars have even bigger corrections. Tau Canis Majoris, to the eye a modest fourth magnitude star in Orion's canine companion, is an O9 supergiant with an absolute visual magnitude of −7.0, but bolometrically it lies at −10.1, not much fainter than Cygnus OB2 #12. The hot O5 supergiant Zeta Puppis, while in absolute terms fainter to the eye than Tau Canis Majoris, shines just as brightly as the latter when the whole spectrum is used. Topping the list is an otherwise obscure supergiant that hovers just below naked-eye vision, HD 93129A ("HD" from the Henry Draper memorial spectroscopic catalogue). Its absolute visual magnitude of −7, spectral class of O3, and a distance of 11,000 light-years reveal an absolute bolometric magnitude of −12, making it the most luminous star known in our Galaxy, shining – in total energy – 5 million times more brightly than our Sun, just topping Eta Carinae.

High-mass main sequence stars come close to matching the bolometric

Figure 5.2. The classic zodiacal constellation Scorpius is set into a bright part of the Milky Way. The luminous hot B1.5 star Zeta-1 Scorpii is indicated by the lower arrow. The image actually contains four stars close together: Zeta-1 is at the right. Red Antares, which looks fairly dim on this blue-sensitive photographic plate, is indicated by the upper arrow. The globular cluster M4 is just to the right of Antares. [From *An Atlas of the Milky Way*, by F. B. Ross and M. R. Calvert, University of Chicago Press, Chicago, 1934. Copyright Part I 1934 by the University of Chicago. All rights reserved. Published June 1934.]

Figure 5.3 Sirius, the brightest star in the sky (as reckoned by apparent visual magnitude), shines at upper right. The arrow points to the open cluster NGC 2362, which is shown in detail in the inset at upper left. The brightest star in the cluster is the O9 supergiant Tau Canis Majoris. [Large photo from *An Atlas of the Milky Way*, by F. B. Ross and M. R. Calvert, University of Chicago Press, Chicago, 1934 Copyright Part II 1936 by the University of Chicago. All rights reserved. Published September 1936.; *inset*: ESO Southern Sky Survey.]

brilliance of the top O and B supergiants, but suffer the visual indignity of appearing much fainter than they really are. Among the best examples is the brightest star of the "Trapezium" quartet, the class O6 illuminator of the Orion Nebula. At the middle of Orion's sword, even binoculars show a shining ionized cloud of interstellar gas. Though a modest fifth magnitude to the eye, Theta-1 Orionis C has an absolute visual magnitude of $-5$. Its high temperature of 45,000 kelvin requires a correction of some four magnitudes, elevating the star to an absolute bolometric magnitude of $-9$. It is actually more luminous than Deneb.

Hot stars are hardly the exclusive domain of stellar brilliance. The F supergiant star Rho Cassiopiae lies at an absolute visual (and bolometric) magnitude of nearly $-10$, comparable to others in yet cooler classes. Indeed, excluding the remarkably bright stars of classes O and B, the brightest stars across the top of the HR diagram have roughly comparable absolute bolometric magnitudes. As a result, as the temperature drops through classes A, F, K, and M, the diameters of these extreme stars must increase. All supergiants, those that lie across the top of the HR diagram, are large. Deneb, for example, has a diameter 200 times that of the Sun and in our Solar System would stretch to the Earth. Even the main sequence's Theta-1 Orionis C is 10 times the solar diameter. Rho Cassiopeiae, however, is almost 700 times larger than the Sun, and would reach over halfway to Jupiter. At the cool limit, Mu Cephei (Herschel's famous "Garnet Star)" and VV Cephei would approach the orbit of Saturn. These are the kinds of stars that gave rise to the term "supergiant," and represent a new extreme to examine, that of the largest stars. The remainder of this chapter will confine itself to the warmer bright stars and leave the largest for the next.

Imagine our planet orbiting such suns. To be properly warmed by the dim red dwarf star LHS 2924, we would have to be only a million kilometers away and our year would be less than an Earth day. Now replace LHS 2924 with Deneb, which (bolometrically) is four billion times brighter. To receive the same amount of heat we do now, the Earth would have to orbit at a distance of 450 astronomical units, 15 times farther than Neptune is from the Sun. Deneb (from its luminosity and theory) has a mass about 25 times that of the Sun. From Kepler's laws of planetary motion, our hypothetical planet would take 2000 years to orbit. We would live our entire lives within one season of the year. Imagine a winter 500 of our years long, a period that would seem short if we were to orbit a star like Cygnus OB2 #12.

## Where are they?

Look at the photographs of the supergiants presented here and at their constellations of residence: Cepheus, Cassiopeia, Cygnus, Scorpius, Carina. Then some night stand outdoors and find them all set into the broad band of light that encircles the sky, the Milky Way. This glowing river, made of the combined light of the stars in the disk of our Galaxy, also contains the gas and dust out of which new stars are forming. Plotting the positions of the brightest stars, of the O and B stars and even the supergiants, we see that they are almost all located within or close to this celestial band, telling that they are within the confines of the our Galaxy's disk. To the naked eye, they fall into a wide circle around the sky called "Gould's belt" (after the nineteenth-century astronomer B. A. Gould), which is somewhat tilted relative to the Milky Way.

More telling, the brightest stars, including the massive O and B stars of upper main sequence as well as supergiants of all classes, have a powerful propensity to congregate together, to form associations, more properly called "OB associations." Dozens are known, ranging in size from that of open clusters like the Pleiades to huge systems over 500 light-years across. They are usually named by their constellation of residence, the first in Cygnus called Cygnus OB1. The stars within them are numbered, the name Cygnus OB2 #12 now making sense. Most of the distances given in the accompanying table are based upon that of the parent OB association, as they nearly all are too far away for parallax measurement.

Several associations are visible to the naked eye. Unlike compact open clusters such as the Pleiades of Taurus, OB associations are so large that we are

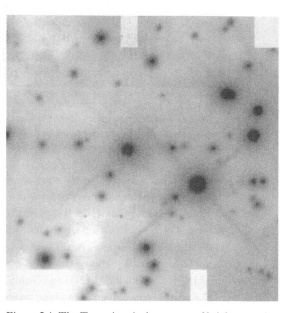

Figure 5.4. The Trapezium is the quartet of bright stars that lights the Orion Nebula (shown here in negative). The brightest one, an O5 "dwarf" with a luminosity 300,000 times that of the Sun, does most of the work. The numerous fainter stars are part of the nebula's dense star cluster, over which the Trapezium reigns. (The Trapezium is seen in context in Chapter 2.) [Calar Alto Observatory, Max Planck Institute for Astronomy, M. J. McCaughrean and J. R. Stauffer.]

Figure 5.5. A photograph of central Perseus made on a blue-sensitive photographic plate dramatically records brilliant blue stars of the Perseus OB3 association (which includes Alpha Persei, the bright star toward upper right) as well as lesser lights. [© National Geographic–Palomar Observatory Sky Survey, reproduced by permission of the California Institute of Technology.]

not immediately aware that the stars are grouped together. The prime examples are nearly whole constellations. Probably the best known is Orion OB1, which consists of Rigel, the somewhat lesser B supergiants Zeta and Epsilon Orionis of the belt, and Kappa, which marks Orion's right leg. Then add the O stars: the third belt member Delta, Lambda of Orion's head, Sigma, and Theta-1 of the Sword, plus numerous lesser lights. Much the same can be said about Scorpius, where Scorpius OB2 contains the B stars Delta, Beta-1, and Tau Scorpii. Associations contain not only stars of the upper main sequence and blue supergiants, but also jewel-like red supergiants, epitomized by Antares in Scorpius (Scorpius OB2), the spatial relationship indicating a physical relationship as well. Perseus is similar.

OB associations have a peculiar property. When we examine their motions by combining their line-of-sight radial velocities with their movements across the sky – their proper motions – the stars of an association are seen to be moving away from each other. The associations are disintegrating and cannot last longer than a few tens of millions of years, a blink on the stellar time-scale (the Sun has a main sequence

Figure 5.6. The vast Lagoon Nebula of Sagittarius is lit by the ultraviolet light of the massive O stars within. [AURA/NOAO/NSF.]

lifetime of 10 billion years). Some of the disintegration is violent. A handful of "runaway stars" are ejected at such high speeds by gravitational interactions or by the explosions of massive companions that they are found even in different constellations, AE Aurigae and Mu Columbae part of Orion OB1 and Zeta Puppis coming from an association in neighboring Vela. Yet for all the movement, a vast number of B stars and nearly all O stars are still within associations. It is obvious both that O and B stars cannot live very long and that they are born together within the confines of the group and quickly die. They must thus be among the youngest stars, defining another extreme category to be explored later.

Associations are linked to the gases of interstellar space, their hot O stars commonly buried within nebulae that are vastly larger than the planetary nebulae. The great Orion Nebula, the archetypal example of a "diffuse nebula," is illuminated principally by Theta-1 Orionis C, which ionizes a hundred solar masses of hydrogen gas out to a distance of over 10 light-years. Mixed in with the nebulosity are countless dark patches of dust that show the gas and tiny solid grains are all mixed together. Scorpius and its neighbor Ophiuchus are loaded with similar, though less obvious, structures. The Milky Way is full of them, including the Lagoon and Trifid Nebulae of Sagittarius, Serpens' Eagle Nebula, the North America and Pelican of Cygnus, and countless more; indeed the diffuse nebulae are *confined* to the Milky Way, as for the most part are the O and B stars and all the OB associations.

The combination of the youth of the O and B stars and the intimate relation between them and the diffuse nebulae powerfully indicate that the gases of interstellar space are the stars' birthplaces and that the diffuse nebulae are the remnants of

Figure 5.7. The spiral arms of the beautiful galaxy M81 are outlined by bright knots made of bright O and B stars and their surrounding diffuse nebulae. [AURA/NOAO/NSF.]

the stuff that made those stars. Red supergiants are commonly found within associations. But it is the O and B stars that are within the diffuse nebulae, showing that they are the initial members of an association. The red supergiants must be the results, the progeny, of the O and B stars. When we look at the stars on an HR diagram, the red supergiants bear the same relation to the O and B stars of the upper main sequence that the giants bear to the lower-mass stars. The beauty of the massive stars is that we can begin to understand the relations of stellar evolution from the observations alone, giving credence to the theories, and, by induction, implying that we are right about the giant stars' descents from red magnificence to white-dwarfhood.

Next, look at the distribution of stars in three dimensions. All are distant. Of the brightest stars listed here, only Rigel, Zeta Puppis, and Theta-1 Orionis C are under 2000 light-years away. These stars – or at least the associations – must be terribly far apart and therefore quite rare. A careful count reveals that 72% of all stars are of the dwarf M variety and that the O stars constitute only an astonishingly low 0.000 04%. Part of the reason is their short lifetimes, but that is not enough: Nature must not like to make really massive stars, tens of times the mass of the Sun. Interstellar clouds far prefer to make low-mass bodies. Why then study these rare beasts? Because they are so bright that they can be seen at immense distances and can be used to chart our Galaxy. They can be observed easily even in fairly distant galaxies, revealing that they lie along graceful spiral arms that our own Galaxy contains as well. Moreover, these massive stars erupt into the devastating explosions of supernovae, which scatter newly-formed heavy atoms into the cosmos, thus powering galactic evolution, and creating some of the matter out of which we ourselves are made.

In extreme instances O stars cluster together into tight, bound groups. Nearby, we see the effect within the Trapezium, a quartet of four O and B stars. Other galaxies have extraordinary systems. The Tarantula Nebula in the Large Magellanic Cloud (a small nearby companion galaxy 150,000 light-years away) is powered largely by a compact cluster called R136 whose core is so densely packed that at one time it was thought to be a "superstar" of over 1000 masses. Over 50 very hot O3 stars and a great number of lesser O stars are stuffed into a volume less than eight light-years across! Imagine living inside such a collection. Your sky would be filled with hundreds of stars brighter than Venus, the brightest rivaling the full Moon, creating a sight almost beyond imagination.

## Spectra, distances, and winds

Rho Cassiopeiae does not look any different from other more common stars, so how do we know it is a supergiant? We can use the distance to find the luminosity or we can simply look at the star's spectrum. The story goes back to 1897, when Antonia Maury invented an extensive spectral classification system that briefly rivalled the simple Harvard scheme of separating the stars into the seven basic classes OBAFGKM. In addition to dividing the stars just by the presence of particular lines (or by ionization, or, as we now know, by temperature) she recognized subclasses for stars with absorption lines that had structural differences. Compared to the average (class *a*), some stars had lines that were fuzzy and broader than usual (class *b*); others (class *c*) had oddly narrow lines, sharp knife-slices taken from the spectrum. The classification was strictly empirical, and though no one at the time knew what it meant, the solutions were not long in coming. In his work that in part led to the establishment of the HR diagram, Ejnar Hertzsprung discovered that the narrow-line *c* stars had very small proper motions; that is they were moving much slower across the sky relative to their neighbors than were the *a* stars. The only conclusion was that on the average subclass *c* had to be much more distant than the others, and consequently vastly more luminous. The *c* designation is all that remains of the Maury system; *a* and *b* (which refer to stars in such rapid rotation that the

Figure 5.8. (See also Plate X.) The Tarantula Nebula, the closest example of a "giant diffuse nebula," lies within a companion galaxy, the Large Magellanic Cloud. It is powered principally by R136 (*in the box, enlarged, at upper right*), a huge cluster of luminous stars packed into a tiny volume. [STScI, NASA, and ESA.]

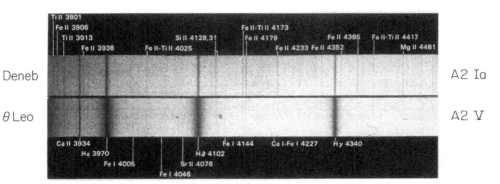

Deneb        A2 Ia

θ Leo        A2 V

Figure 5.9. Supergiant Deneb and the third (apparent) magnitude dwarf Theta Leonis are both stars of class A2, and accordingly have dominating hydrogen (H) lines and similar temperatures. However, the spectrum of supergiant Deneb has much narrower lines, the result of the low density that goes along with being a supergiant. (Roman numerals give ionization states, "I" meaning neutral, etc.) [From *An Atlas of Representative Stellar Spectra*, by Y. Yamashita, K. Nariai, and Y. Norimoto, University of Tokyo Press, 1978.]

absorption lines are blurred by the Doppler effect) are never used. We still hear the term echo in the "Lc" supergiant irregular variable stars as well as in a few other instances. The criterion of narrow lines applies only to hotter stars in classes O, B, and A. However, there are other standards that can be tied to cooler stars. Compared to their strengths in G and K main sequence stars, the cyanogen (CN) molecular bands weaken in giants and even more in supergiants. In each case the difference is a result of the larger stars' lower densities. Modern spectroscopic techniques are sufficiently fine to recognize all the different classes of stars, and even to discriminate among bright and less-bright giants and supergiants.

Just because we can recognize supergiants by their spectra does not mean that we know how bright they are. For that we need actual distances, and all but a few O stars and luminous supergiants are too far away – in spite of recent advances in observational techniques – for the measurement of distance by direct trigonometric parallax. Instead, we make use of clusters. Clusters have great importance as tools in the study of stellar evolution and the Galaxy, since all the stars within one were born at about the same time. As direct evidence, clusters containing O and B stars are commonly allied with youthful OB associations and with diffuse nebulae. Most clusters, however, do not contain O and B stars. Since we already know that these stars do not live very long lives, those absent of O stars must be older. There is a continuum of

Figure 5.10. (See also Plate XI.)The young Double cluster in Perseus, within the great Perseus OB1 association, itself contains O stars as well as red supergiants. [M. E. Killion.]

clusters in which progressively more of the upper main sequence is missing. All clusters, however, have lower main sequences that consist of G, K, and M stars. These stars live the longest and must have ages that could be as great as that of the Galaxy itself.

The continuum of clusters empirically demonstrates that not only do the massive O stars live the shortest time, but that potential lifetimes progressively increase as we descend the main sequence. To explain what we see, upper main sequence stars must live shorter periods of time, consistent with the demonstrable youth of OB associations. In the beginning a cluster must display no more than a main sequence (the zone of stable hydrogen fusion), along which are distributed stars of all masses, from the lower limit at 0.08 that of the Sun to perhaps 100 times solar. The higher the mass the quicker must be the pace of evolution, and the faster the stars die. A cluster loses its main sequence from the top down, high-mass stars first becoming luminous supergiants (also found in young clusters), and those of lower mass transforming themselves into more ordinary giants, which are found in older clusters like Cancer's Beehive and Taurus's Hyades.

Modern parallax techniques allow astronomers to measure the distances to several nearby clusters, including both the Hyades and the Pleiades, respectively 150 and 425 light-years away, which together contain stars up the main sequence to class B as well as several kinds of giants. The known distances give the absolute magnitudes of the clusters' stars and allow the establishment of a combined HR diagram, albeit without O stars and supergiants. We next establish an HR diagram for a young cluster (the Double cluster perhaps) in which *apparent* visual magnitude is plotted against spectral type, then overlay its diagram on top of that of the combined–cluster

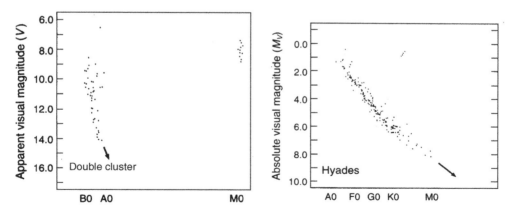

Figure 5.11. Clusters lose their stars from the top down. The main sequence of the combined Double cluster (*left*) extends to the O stars, whereas that of the Hyades (*right*) stops short in class A. The Double cluster contains supergiants, whereas that of the Hyades only giants. The Double cluster must therefore be younger than the Hyades. [Adapted from *An Atlas of Open Cluster Colors–Magnitude Diagrams*, by G. C. Hagen, Publ. of David Dunlap Observatory, Toronto, 1970. (Art from *Astronomy!: A Brief Edition*, J. B. Kaler, © 1997, used by permission of Addison Wesley Educational Publishers Inc.)]

diagram plotted with *absolute* visual magnitude, line up the spectral classes on the lower axis, and slide the top one up and down until the main sequences fit. Not only is the main sequence extended upwards to higher masses, but the difference between absolute and apparent magnitude for the young cluster is found, from which the actual distance is easily calculated. Enough clusters subjected to such treatment give us absolute magnitudes of all kinds of stars, including the most luminous.

More important, the technique, combined with parallaxes, allows the correlation of spectroscopic features (line widths for example) with absolute visual magnitude, further allowing the recognition of any particular kind of star from its spectrum and the determination of its distance. The only kinds of stars for which the method fails are those with peculiar spectra that have yet to be calibrated in terms of absolute magnitude, like those that inhabit the planetary nebulae, as only a tiny handful are in clusters. We must then resort to indirect methods that are not very accurate.

In the 1940s, William Morgan of Yerkes Observatory and Philip Keenan of Ohio State developed the now-standard classification for stellar luminosities and evolutionary stages in which Roman numerals I through V denote supergiants, bright giants, giants, subgiants, and main sequence dwarfs. The Sun is a G2 V star, variable Mira averages M7 III, and Rigel is B8 I. The supergiants, however, occupy an enormous range in luminosity. To describe them, Morgan and Keenan broke the class in two, calling the brightest supergiants Ia and the lesser versions Ib. (Rigel is a B8 Ia star). Of the exemplary stars in the table of supergiants included here, only Tau Canis Majoris is of the Ib variety.

Yet division into Ia and Ib is still insufficient. Keenan suggested that even the term "supergiant" could not convey the true grandeur of the very brightest – and rarest – of stars, which are called by the lofty name "hypergiant." Here the Roman numeral classes break down, as the Romans had no zero, so Keenan appropriated the Arabic zero, 0, to describe the hypergiants' luminosities, leading to some confusion with the decimalization of the spectral classes (not to mention letter-class O). The "zero" class was first applied to the brightest stars of the Large Magellanic Cloud, of which the massive S Doradus is an example (classified A5 0). In our Galaxy, Eta Carinae (B0 0), a hypergiant with a mass of some 100 suns, qualifies. Several others (like P Cygni) of somewhat lesser luminosity fall between the supergiants and hypergiants and are called "Ia-0.

Of the 19 stars in the accompanying table (p. 101), seven have an *e* or an *f* appended to their formal class, telling that the spectrum contains emission lines in addition to the usual absorption features. These emissions reveal a star to be surrounded by a cloud of low-density gas, by matter lost through a strong wind, as are so many of the M giants. And, as in the case of the cool giants, the stars can change their surface compositions. Among the O stars *e* indicates hydrogen-line emission; but *f* denotes emission in both helium and nitrogen. The O stars, by definition, have strong ionized helium absorption: the hydrogen lines are weaker than they are in classes A and B in part because in the high-temperature environment the hydrogen becomes ionized and there are fewer neutral atoms to absorb the radiation (a change in the transparency of the spectrum-producing stellar atmosphere also contributes). Among the A stars, hydrogen lines narrow as luminosity increases. As luminosity climbs, the helium absorptions in the O stars narrow and weaken as well, but the effect is much more extreme. At some point the principal line of ionized helium at 4686 ångstroms actually disappears, and for the most luminous stars reverses to become an emission feature. At the same time several nitrogen lines become prominent in emission as well, partly in response to a luminosity-type effect, and partly because the surface nitrogen abundance probably increases as a result of interior evolution, mass loss, and convection.

Mass-loss rates can be remarkably high: Zeta-1 Scorpii's wind is measured to

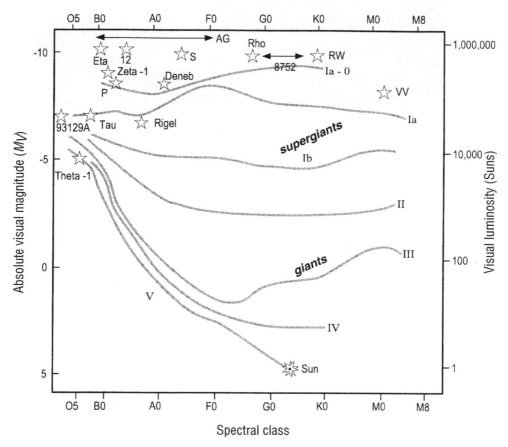

Figure 5.12. Supergiants are subdivided into subclasses Ia and Ib, and hypergiants (called 0) are even brighter. The most luminous stars are plotted with abbreviated names. The relative faintness of the stars hotter than B0 is a consequence of having so much of their light emitted in the ultraviolet. Two stars can occupy the same position and yet be very different. P Cygni is in a very different evolutionary state than Zeta-1 Scorpii. [Adapted from a diagram in the author's *Stars and their Spectra*, Cambridge University Press, 1989.]

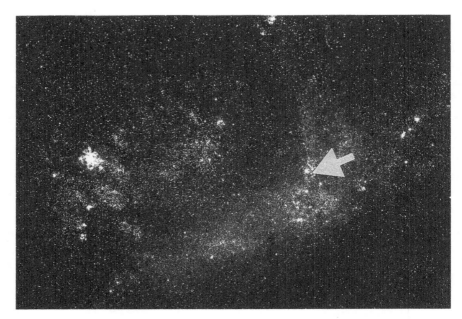

Figure 5.13. The Large Magellanic Cloud, a low-mass companion to the Milky Way, has a very high star formation rate and contains numerous hypergiants including S Doradus (*arrow*). The great Tarantula Nebula is seen in context at left. [AURA/NOAO/NSF.]

blow away some five hundred thousandths of a solar mass per year, almost a billion times the rate of mass loss in the solar wind, and comparable to the greatest mass-loss rates found among the giant stars. Given the lifetimes of stars, these rates mean that an individual can eventually lose many solar masses, cutting itself to a fraction – perhaps half or less – of its original self, a fact that will profoundly change its appearance. Some stars even have visible nebulosity around them, gas that has been blown to large distances and then illuminated by the ultraviolet radiation of the hot stellar surface.

A leader among windy hot stars is P Cygni, a remarkable body that is evaporating at a vigorous rate of four ten-thousandths or so of a solar mass per year. The outflowing wind is so dense that it produces the peculiar looking spectrum lines described earlier, those that have absorptions to the short-wave sides of emission lines, a phenomenon that indicates fierce mass loss. P Cygni, visible to the naked eye, was the first one in which such spectrum lines were found, and has given them their name ("P Cygni lines") that memorialize one of the grandest stars of the Galaxy.

The brightest of the hot supergiants commonly show such lines. In the spectrum of the brilliant second magnitude O4 If star Zeta Puppis, the most powerful feature in the ultraviolet comes from triply-ionized carbon. The left-hand absorption edge is measured to have a wavelength 12 ångstroms shorter than it would be were the gas at rest, indicating an outflow velocity of 2300 kilometers per second. Though not as great as the speed in the winds of planetary nuclei, the amount of matter within a massive supergiant's wind is vastly greater than in their little counterparts. Zeta Puppis loses matter at a rate of over a millionth of a solar mass per year, 10 million times the flow rate of the solar wind. At the extreme, such winds can impact the local environments, OB associations helping (along with exploding stars) to blow bubbles within the surrounding dusty interstellar gases.

## Extraordinary variation

Stellar variation – variation in brightness – is found among all kinds of stars. Most giants shine quietly. But the large dimensions of luminous, advanced-evolution AGB giants, coupled with their internal structures, lead to the dramatically variable Mira stars. Smaller stars are generally more stable, but here and there on the lower main sequence are stars whose magnetic stresses produce sudden and surprisingly bright flares. Even the nuclei of a few planetary nebulae are subtly variable, a subject to be examined under the smallest stars (to which the planetary nebula stars also belong).

Hot bright stars, those on the upper main sequence, do not in general undergo notable fluctuations, nor do the common warmer supergiants. The stars of Orion's Trapezium stare down serenely, and night after night Deneb and Rigel will appear quite constant to the eye. (Deneb does indeed show some variability, but only in its spectrum lines.) Among the most luminous stars, obvious variability is reserved mostly for the large stars of the cool classes, the M supergiants. There are distinct exceptions, however, among the hypergiants, which produce some of the most wondrous stars of the Galaxy.

Among the greatest of these is Eta Carinae, the centerpiece of the vast nebula in the southern hemisphere that shares its name. This awesome cloud of dusty gas, visible to the naked eye, marks a region rich in massive stars. During most of the 18th and 19th centuries, Eta Carinae, over 8000 light-years away, shone between second and fourth apparent magnitude. Around 1840, it began to brighten, and by 1848 had reached apparent magnitude $-1$. At that time the sky's three brightest stars lay along a graceful 55°-long arc, with Sirius at the northern end, Eta at the southern, and Canopus in the middle. Eta then began a great drop, and by 1880 had plummeted below naked-eye visibility to 8th magnitude. A primitive spectrogram taken in 1893 shows that it had the spectrum of an F supergiant. The star has since climbed to 6th magnitude, and the stellar spectrum has disappeared to be replaced by the emission spectrum of a surrounding dusty nebula. The star itself, so involved in its bright tiny

Figure 5.14. In the spectrum of P Cygni, the emission lines (bright) are flanked to shorter wavelengths by absorptions (dark). The three strongest lines are from hydrogen, the next brightest from neutral helium. [From *An Atlas of Representative Stellar Spectra*, by Y. Yamashita, K. Nariai, and Y. Norimoto, University of Tokyo Press, 1978.]

Figure 5.15. The great Eta Carinae Nebula dominates the southern Milky Way. Eta itself (*arrow*), a *luminous blue variable*, was a first magnitude star in the last century and now is at sixth; HD 93129A, the brightest star in the prominent cluster toward upper right, is the most luminous star known in the Galaxy. [AURA/NOAO/NSF.]

nebula, is difficult to see. Infrared observations, which can penetrate the dust, reveal four compact objects, the star and three other small blobs that for a time were thought to be stellar companions that would make a system reminiscent of Orion's Trapezium.

Hubble Space Telescope observations zero in on the star's real nature. The telescope's position above the Earth's disturbing atmosphere allows it to separate the star from the surrounding cloud of dusty gas. What we see is not quite the star itself, but the star illuminating an outflowing wind, one of the most powerful known. The luminosity of 4.5 million Suns – corresponding to an absolute bolometric magnitude of −12 – shows it to be a hypergiant. Analysis of the spectrum indicates a stellar temperature between 20,000 and 30,000 kelvin and a spectral class between B0 and B1 with a current mass outflow rate in the wind of a thousandth of a solar mass per year, a hundred times that of an advanced Mira variable. The small objects near it are not stars at all, but tiny blobs of matter that were expelled by the star in an episode of even greater mass loss.

Other Hubble observations show the nebula about the star to consist of an immense bipolar flow, a double-sided flow of gas away from the star that has produced a bipolar nebula vaguely reminiscent of the bipolar planetary nebulae but on

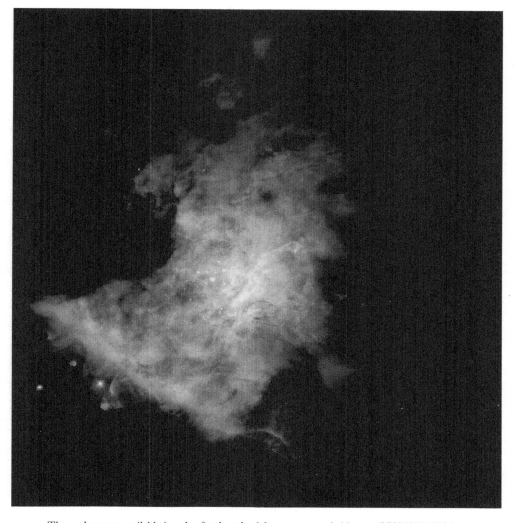

I  The great Orion Nebula, viewed here by the Hubble Space Telescope, is lit by the four hot stars in the Trapezium at its center, the group dominated by Theta-1 Orionis C.

II  The brown dwarf Gliese 229B glimmers red next to its vastly brighter M1 dwarf companion. At a temperature of only about 1000 degrees kelvin, its infrared spectrum contains absorption bands of methane, which are never found in stars.

III  Above is a peaceful sunset that might, when the Sun is brightening on its first ascent of the giant branch, look something like the scene below.

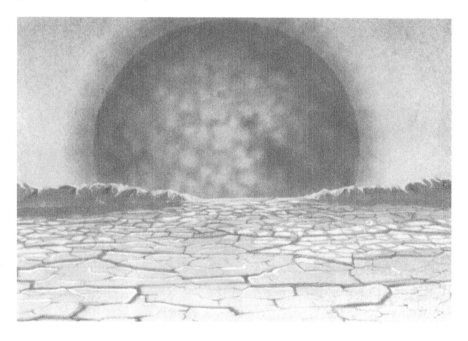

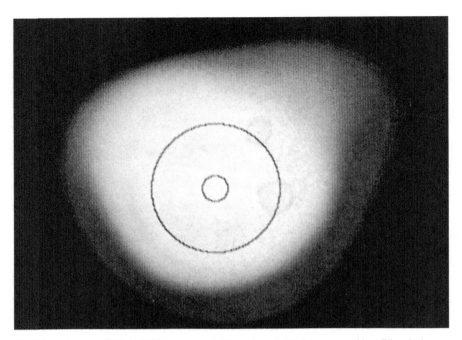

IV The carbon star IRC+10216 is surrounded by a dust shell of its own making. The circles indicate the position of the star and the inner edge of the dust shell.

V Different molecules appear in different locations in the ejected shell around the carbon star IRC+10216, each dependent on different conditions and the degree of illumination by starlight. Each map is 6000 AU across. From inside out we see hydrogen cyanide (HCN), cyanogen (CN), cyanoacetylene (HC$_3$N), and unstable cyanoethynyl (C$_3$N).

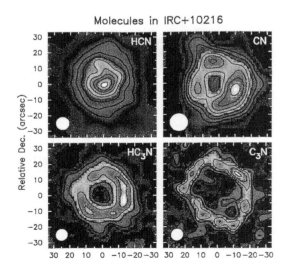

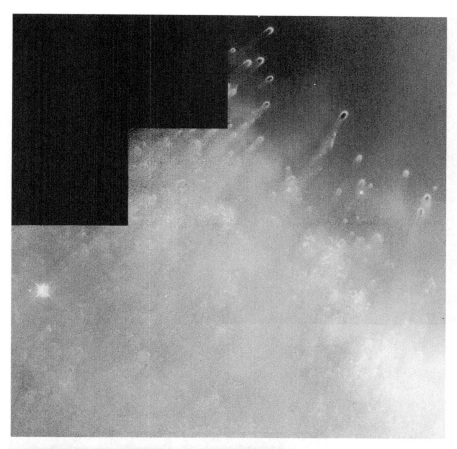

VI (*above*) The Helix Nebula's closeness together with the Hubble Space Telescope allows us to see the details of the extraordinary knots of matter in the central hole that blend to form the inner ring.

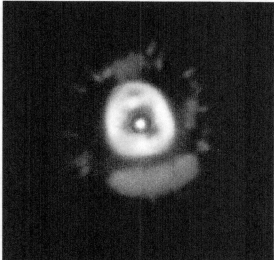

VII This Hubble Space Telescope view of the Eskimo Nebula, NGC 2392, a planetary nebula in Gemini, reveals spectacular complexity. The central star, from which came the expanding gaseous cloud, has a temperature of 70,000 degrees kelvin, and is still heating.

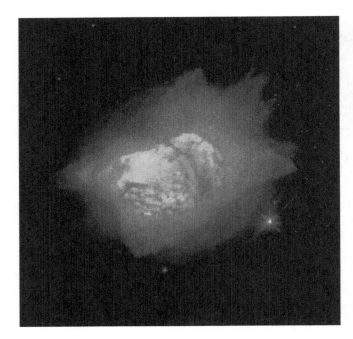

VIII  NGC 7027 is a complex, dusty nebula that has one of the highest central star temperatures known, so high that very little of the star's light is radiated in the optical. The nebula is rich in carbon, implying that its progenitor was an AGB carbon star. The concentric shells represent earlier episodes of mass loss.

IX  The Egg Nebula in Cygnus, observed by the Hubble Space Telescope, consists of dozens of expanding dusty shells. Light from a warm star hidden in a thick disk illuminates the surroundings through the disk's poles. Eventually, the star's wind will compress its surroundings to a shell that will be ionized as the star heats to a temperature beyond 25,000 K.

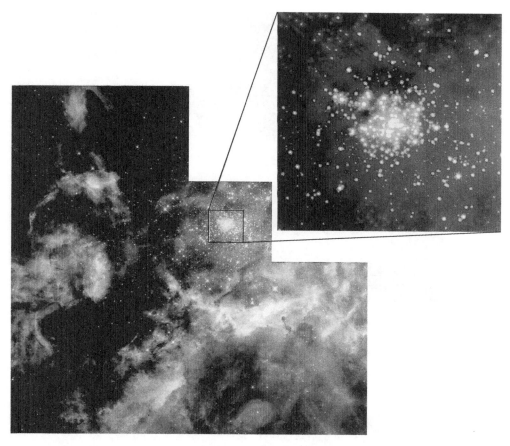

X  The Tarantula Nebula, the closest example of a "giant diffuse nebula," lies within a companion galaxy, the Large Magellanic Cloud. It is powered principally by R136 (*in the box, enlarged, at the upper right*), a huge cluster of luminous stars packed into a tiny volume.

XI  The young Double cluster in Perseus, within the great Perseus OB1 association, itself contains O stars as well as red supergiants.

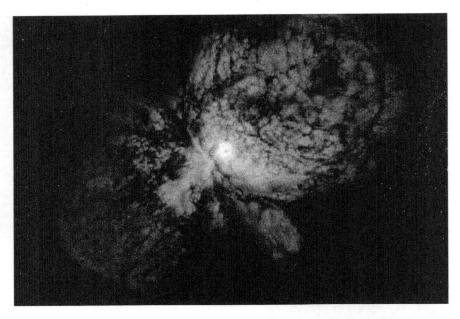

XII A spectacular Hubble Space Telescope view of Eta Carinae shows a pair of bubbles, a bipolar flow presumably blown along the massive star's rotation axis during its enormous outburst over a century ago. Black lanes of dust thread through the illuminated gas.

XIII The Hubble Space Telescope shows the luminous blue variable AG Carinae (the star hidden behind the dark disk) to be surrounded by a knotty irregular nebula.

XIV The Hubble Space Telescope's near-infrared camera was needed to punch through the thick dust of our Galaxy's disk to see the so-called "Pistol Star," a "luminous blue variable" like Eta Carinae. Located close to the center of the Galaxy, its surrounding ejecta stretch over 4 light-years.

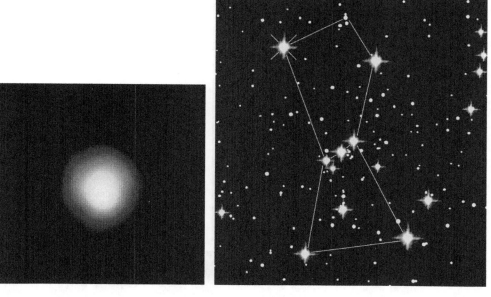

XV An image of Betelgeuse (upper left in Orion) taken with the Hubble Space Telescope shows the star's disk and a bubble of hot gas on its surface.

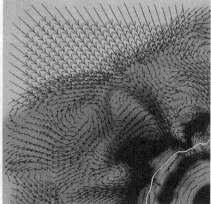

XVI Supernova 1987A, the aftermath of the evolution of a blue supergiant, exploded in the Large Magellanic Cloud. The supernova is the bright "star" toward the lower right. The bright, diffuse object toward upper left is the Tarantula Nebula.

XVII A computer simulation shows a quarter view of the turbulent core of a supernova only 120 kilometers across, a few-hundredths of a second after the collapse. The wavy white line is the wall of outwardly pushing neutrinos. A neutron star is being created within.

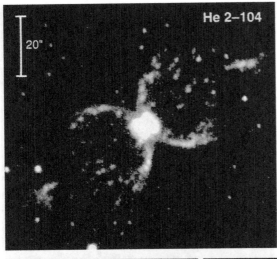

He 2-104

20"

XVIII (*left*) The "Southern Crab" is a planetary nebula with a symbiotic star at its center, and is the result of mass loss stirred by a hot white dwarf.

XIX (*below*) The pulsar in the Crab Nebula, here radiating X-rays, turns on and off 30 times per second. The flash lasts for only a few-thousandths of a second.

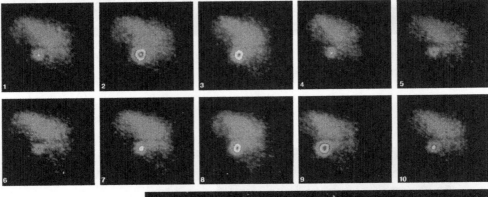

XX The Hubble Space Telescope reveals a tiny non-pulsing neutron star (*arrow*) radiating by virtue of its 1.2 million degrees kelvin temperature.

XXI Earth and Sun are partners, born at nearly the same time, 4.5 billion years ago.

XXII FU Orionis, a massively brightened T Tauri star, is surrounded by a large dusty cloud.

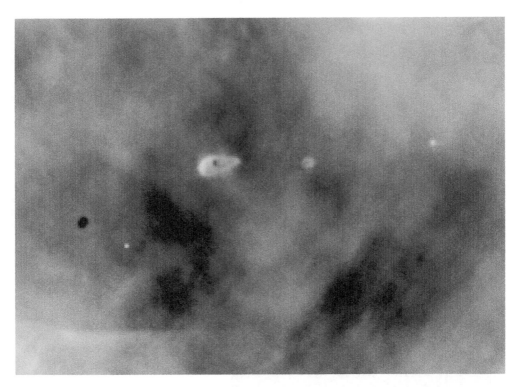

XXIII Associated with the Orion Nebula are numerous disks surrounding young stars, most lit by the bright Trapezium stars that power the nebula. One dark disk to the left is seen in relief against the bright gas.

XXIV The biggest of the terrestrial planets, Earth, is contrasted with the Moon. The other terrestrials, Mercury, Venus and Mars, are similarly structured, and consist largely of iron and rock. The apparent proximity of the Moon to the Earth is the result of foreshortening.

XXV Jupiter, king of the planets, incorporates light stuff, and is made mostly of hydrogen and helium. Its outer large satellites are largely water ice.

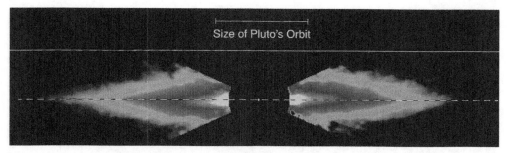

XXVI A dusty disk, 800 AU wide, surrounds the class A main sequence star Beta Pictoris. Evidence that the disk may contain planets includes the chemical composition of the dust, the warping of the disk, and an interior hole where planets may have been accumulated from the dust. (The "hole" seen here is produced by an occulting device that hides the bright star.)

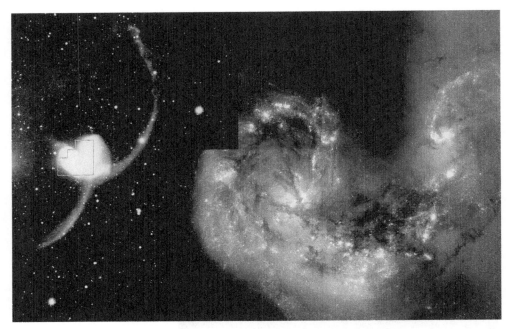

XXVII The "Ring-Tail galaxy" in Corvus consists of two colliding galaxies, NGC 4038 and NGC 4039, that send out long tidal streamers (*left*). A close-up of the interior taken by the Hubble Space Telescope (*right*) reveals hosts of new stars forming within the maelstrom of the collision.

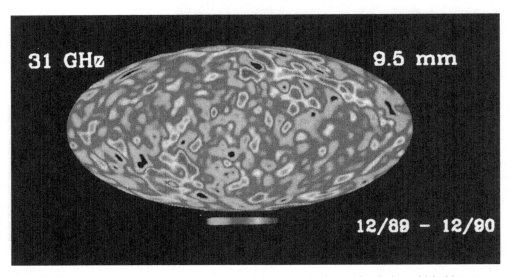

XXVIII This whole-sky map shows the brightness of the cosmic background radiation, which shines at a frigid three degrees above absolute zero, as observed by the *COBE* (= Cosmic Background Explorer) satellite. Tiny fluctuations of only a millionth of a degree reflect variations in density from which galaxies and their clusters ultimately arose.

XXIX The Hubble Deep Field, a 100-hour exposure with the Hubble Space Telescope, reveals galaxies of all kinds terribly far away, to billions of light-years, allowing us to look billions of years back into the past.

XXX The Calabash Nebula, named after its odd gourd-like shape, surrounds a long-period M9 Mira that lies at the dense neck of the object where the two bubbles meet. The star is optically hidden from view by a thick disk of dust; gas streams out through the poles producing shock waves as it batters the surrounding medium.

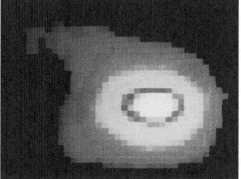

XXXI An infrared view of R Coronae Borealis shows a huge cloud of glowing dust several minutes of arc and over a light-year across that surrounds the star proper. The cloud is the dusty remains of many episodes of the mass loss that have produced the random drops in visual brightness.

XXXII Home, our solid Earth, made mostly of the debris of supernovae.

XXXIII The grandeur of a living rose places the Sun among the true wonders of the Universe.

a vastly grander scale. Perpendicular to the flow is a "waist-band" of stuff presumably in the star's equatorial plane, the bipolar flow directed along the star's rotational axis. Enmeshed within the flow is a network of filaments of dark dust. The three blobs are most likely part of the equatorial flow, the whole bipolar structure over half a light-year in extent.

From these observations we can picture what must have happened. A century-and-a-half ago, the star erupted in a wind whose flow was 100 times the present rate; over the eruption period it may have lost a solar mass of matter or more. The matter was lost most thickly at the star's equator; the rest then could escape at a faster rate along the polar axis where there was less resistance. As the gas hit the colder temperatures of interstellar space, some of it condensed into dust to form the dirty cloud now seen. The star then buried itself in its own effluvia, and dimmed to the eye. For reasons no one understands, matter is not lost symmetrically and smoothly, but at least on occasion in discrete blobs.

However, all may be not as seems. The star's emission lines shift periodically over a period of 5.5 years, suggesting that indeed there is a binary companion. Moreover, the nebula is rich in nitrogen, but the visible star is not. The most reasonable explanation is that the great eruption and the nebula were produced not by the star we see but by the companion. To have ejected so much nitrogen, the companion must be much more evolved, having lost a good portion of its outer envelope to reveal changes wrought by nuclear fusion. More-massive stars live for shorter periods of time. To be more evolved, the companion therefore had once to be more massive than the star we now see. From its luminosity, the visible star must have started with a mass of around 100 Suns, so the two together easily topped 200! After losing mass through winds, the visible star has probably lost 30 solar masses, the now-dimmer (or hidden) companion closer to 40. The lost mass alone dwarfs not just the Sun but the stars of the nearby solar neighborhood out to several light-years!

Whether one star or two, Eta Carinae – bright and intrinsically hot and blue, though buried – is among the Galaxy's greatest "luminous blue variables," or "LBVs." Such stars, while ordinarily quite windy anyway, are characterized by sudden vast mass outflow events that take place centuries apart. The stars actually change their total luminosities – their bolometric magnitudes – rather little. But the encompassing wind absorbs and blocks the starlight, re-radiating it at different wavelengths. Invisible ultraviolet energy may suddenly shift and emerge in the

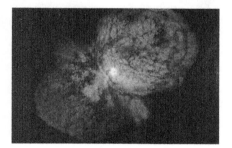

Figure 5.16. (See also Plate XII) A spectacular Hubble Space Telescope view of Eta Carinae shows a pair of bubbles, a bipolar flow presumably blown along the massive star's rotation axis during its enormous outburst over a century ago. Black lanes of dust thread through the illuminated gas. [K. Davidson (University of Minnesota) *et al.*, STScI, and NASA.]

optical spectrum, making the star appear to brighten; or the dust may re-radiate in the infrared, making the star appear to dim, as it has Eta Carinae. The cloud that absorbed the light for a while acted as a false, cooler stellar "surface," and at the end of the nineteenth century made the B supergiant look like a cooler F star.

Such stars, at the upper limit not just of stellar luminosity but of stellar mass, are exceedingly rare. Yet Eta Carinae is not alone. P cygni, the star that gave its very name to spectral features that typify mass loss, enormously brightened and was recorded as a "nova" in the year 1600. Real novae are the result of explosions caused by the transfer of mass from a main sequence star to a white dwarf companion (events to be covered along with the smallest stars). We know now that the P Cygni event likely had some relation to the one involving Eta Carinae, allowing us to recognize the star as an LBV. Oddly in the same constellation as Eta Carinae, we find AG Carinae, also surrounded by a cloud of ejecta that has seemingly come off in irregular blobs. No one knows why stars eject matter in this erratic fashion, although the irregular shapes of the Mira variables may supply a hint. Whether a binary companion is involved, as it may be for Eta Carinae, is unknown.

Another, though probably lesser example, is Rho Cassiopeiae. Ordinarily a relatively cool supergiant of class F, in 1945 it ejected a huge cloud and to the eye dropped to class M, as it expelled matter that acted as a false, cool stellar "surface." Among the grandest examples, certainly one that rivals Eta Carinae, is the so-called "Pistol Star," recently discovered with the Hubble Space Telescope near the center of the Galaxy. Astronomers have suggested an initial mass as high as 200 times that of the Sun.

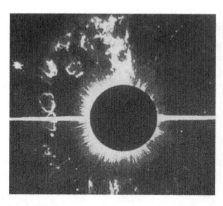

Figure 5.17. (See also Plate XIII) The Hubble Space Telescope shows the luminous blue variable AG Carinae (the star hidden behind the dark disk) to be surrounded by a knotty irregular nebula. [A. Nota, STScI, and NASA.]

Figure 5.18. (See also Plate XIV) The Hubble Space Telescope's near-infrared camera was needed to punch through the thick dust of our Galaxy's disk to see the so-called "Pistol Star," a "luminous blue variable" like Eta Carinae. Located close to the center of the Galaxy, its surrounding ejecta stretch over 4 light-years. [D. F. Figer (UCLA), STScI, and NASA.]

Other galaxies offer more. S Doradus, one of the most luminous stars of our most notable companion galaxy, the Large Magellanic Cloud, provides a wonderful example. Though 160,000 light-years away, the star is of the eighth magnitude and visible in binoculars. Then, 35 years ago, Edwin Hubble and Alan Sandage found a few immensely luminous variables in the Triangulum Spiral M33, called just "Variables A, B, C and 2." They are faint to us on Earth, only 18th magnitude or so, but are among the brightest in this nearby galaxy, which is 2.5 million light-years away. These Hubble–Sandage variables appear to be hot supergiants with shrouds that again mimic the spectra of cooler stars. Variable C has decade-long episodes of brightening about every 40 years or so; variable A, rather like Rho Cassiopeiae, turned from an F to an M star in just a year, dimming by 3.5 magnitudes in the process. They too are LBVs.

Whatever their outward forms, in the centers of the brightest LBV systems beat fiercely bright blue hearts – some possibly double – that produce so much energy that the stars are tearing themselves apart. One origin for such fierce mass loss is simple radiation pressure. A star like the Sun is stable because at any point the inward pull of gravity is just balanced by the outward pressures produced by the hot gas and by the outflowing radiation that is absorbed (and re-radiated) by the gas. In the Sun, the contribution of radiation pressure is very low. However, as the mass of a star increases so does its central temperature, the rate at which nuclear fuel is burned, and the luminosity. At some point, at the dangerous "Eddington limit" (after Sir Arthur Eddington), the amount of radiation becomes so high that its pressure exceeds the ability of gravity to hold the star together. The LBVs are near the upper limits of what is allowed, and are evaporating because of it, but, for unknown reasons, not smoothly but in episodic fountains. Gravitational action by companions may be involved as well, as perhaps are other mechanisms. The stars remain wonderfully mysterious.

## Wolf–Rayet stars

Named after the French astronomers Charles Wolf and Georges Rayet, who discovered them in 1867, the Wolf–Rayet stars are as strange as the LBVs.

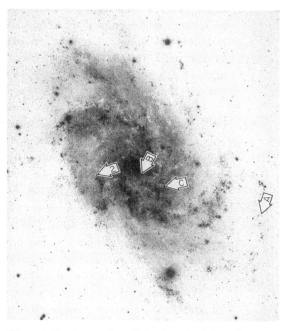

Figure 5.19. The nearby spiral galaxy M33 (seen here as a negative image) displays luminous blue variables labelled A, B, C, and 2. These stars are easily seen, though 2.5 million light-years distant. [Palomar Observatory, California Institute of Technology.]

119

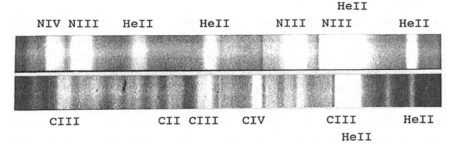

Figure 5.20. The blue-violet spectra of two Wolf–Rayet stars stretch from around 4000 Å to 4700 Å, nitrogen-rich WN6 HD 192163 (*top*) and carbon-rich WC7 HD 192103 (*bottom*). Helium, the dominant element, is seen in both. [From *An Atlas of Representative Stellar Spectra* by Y. Yamashita, K. Nariai, and Y. Norimoto, University of Tokyo Press, 1978.]

They are luminous hot supergiants with temperatures comparable to those of the normal O stars. However they cannot actually be placed in this class because of their eccentric spectra, which display only emission lines and little or no evidence of hydrogen, normally the most common of elements. Their luminosities range between about 100,000 and a million times that of the Sun, at the limit close to those of the LBVs. Though rare (there are probably only 1000 or so in the Galaxy), they are at least more common than the LBVs. Gamma-2 Velorum, one of the sky's brighter stars, shining at apparent visual magnitude 1.8, is a double comprising an O giant and a Wolf–Rayet star. Also like the LBVs, the Wolf–Rayet stars are losing mass, at high rates, a ten-thousandth to a hundred-thousandth or so of a solar mass per year, in which the dominant element is not hydrogen, but helium.

Wolf–Rayet stars come in two flavors, nitrogen-rich (WN) and carbon-rich (WC). WN stars do contain some hydrogen, although the ratio of hydrogen to helium is reversed. In normal stars, there is about 10 times as much hydrogen as helium, whereas in the WNs, there is typically 3 to 10 times as much helium as hydrogen. While carbon and oxygen are effectively absent, WNs contain up to 10 times as much nitrogen relative to helium (and vastly more relative to hydrogen) as does the Sun. If the suspected companion to Eta Carinae is not already a nitrogen-enriched WN star, it will be soon. The WCs are more extreme. In these, no hydrogen is seen at all and neither is nitrogen, while the ratio of carbon to helium is 100 times normal. In the extreme, the number of carbon atoms nearly equals that of helium. Here we find oxygen somewhat elevated relative to helium.

Within each category there is a range of properties, particularly in the level of ionization of nitrogen and carbon, which apparently correlates with a stellar temperature that runs from 50,000 K to perhaps as high as 100,000 K. In addition, the carbon abundance of the WC stars seems to increase with increasing temperature. The Wolf–Rayet component of Gamma-2 Velorum, the bright naked-eye Wolf–Rayet star, is one of the cooler WC stars. A third group, the WO stars, is like WC but with a higher oxygen level.

The stars have apparently removed their outer hydrogen envelopes to expose layers that are rich in the by-products of different kinds of nuclear-burning, allowing us to see the aftermaths of nuclear furnaces right before our eyes. The nitrogen-rich version is showing the results of the carbon cycle, which while making helium from hydrogen, also manufactures fresh nitrogen from carbon. Though the nitrogen turns into oxygen (and then back to carbon), the cycle upsets the normal balance, increasing the ratio of nitrogen to carbon to well above the solar value (hence the nitrogen enrichment of the nebula around Eta Carinae). The richness of the carbon stars, on the other hand, can be produced only by the fusion of three atoms of helium into one of carbon, the fusion chain that powers the giant stars and that also obviously plays a powerful role in running the supergiants. How the different stars relate to each other and to the pathways of stellar evolution is not known, though it is suspected that WN stars may turn into WC stars.

The great amounts of matter lost from such stars sometimes appear as grand nebulae, ionized and illuminated by the hot stars within. Such nebulae can look very much like planetary nebulae, and are really caused by much the same processes, the wind from the Wolf–Rayet stars slamming into the mass that was lost while the stars were in the process of coming into their present states. These nebulae, however, are much more massive than (at least most) planetary nebulae, and to distinguish them from their lesser siblings, they are – from their appearances – called "ring nebulae" (not to be confused with the "Ring Nebula in Lyra," a true planetary nebulae derived from a lower-mass star). Ring nebulae are also much more enriched with the by-products of thermonuclear fusion than are planetary nebulae, and contain the same elements that show up in the surface atmospheres of their parent stars: ring nebulae surrounding WN stars are rich in nitrogen and those surrounding WC stars rich in carbon. As are the planetary nebulae, these stars are providing a load of heavy elements for future generations of stars to process – and perhaps to make planets as well.

Wolf–Rayet stars give testimony to the huge amounts of matter that stars can lose through their winds. They commonly occur in double systems that allow their masses to be determined. Though a Wolf–Rayet star contains in the neighborhood of 20 solar masses, a main sequence companion can weigh considerably more. To have evolved first, however, the Wolf–Rayet component must once have been the more massive of the two, the star losing perhaps tens of solar masses. The suspected companion of Eta Carinae may be doing that even as we watch.

Wolf–Rayet stars may be the leavings of luminous blue variables that have finally lost their outer hydrogen envelopes and bared their cores. Their probable fates, like those of all massive stars, is to explode, to produce supernovae and the stellar extrema known as neutron stars, or even black holes, something that more modest stars like the Sun cannot do. The most luminous stars are very different in ways that involve much more than brightness. We will see how later, after first examining the behemoths of the Galaxy, the great red supergiants.

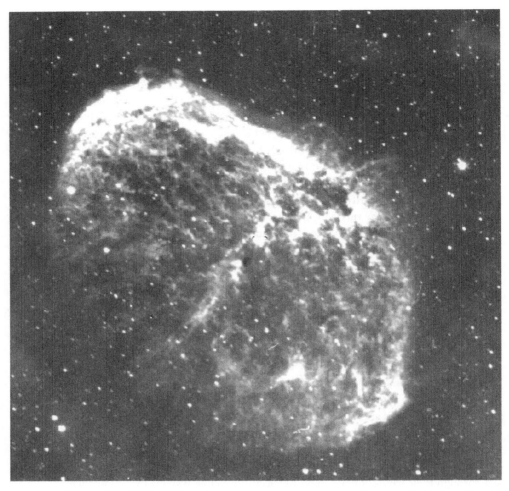

Figure 5.21. NGC 6888, in Cygnus, is a complex, nitrogen-rich ring nebula ejected by the Wolf–Rayet star at the center. [Case Western Reserve University Burrell Schmidt, K. B. Kwitter.]

# Chapter 6

# The largest stars

Move now to the upper right-hand corner of the HR diagram and examine the largest stars, joining them with the two neighboring sets of stellar extrema, the brightest and coolest stars. No matter how long one studies astronomy the range of stellar sizes remains astonishing. The smallest stars, the remnants of supernova explosions, are no bigger than a small city; the largest stars, extreme supergiants and the stars that produce some of the eruptions that in turn make the smallest stars, are comparable to the size of the planetary system.

## The characters

While not at the very top of the brightness scale, the largest stars are still among the most luminous. Look at Orion. As he pursues his hunt across northern hemisphere winter skies, he carries stars that help define the broad variety of supergiants and that lead us to the realm of the largest stars. The constellation is dramatically outlined by two magnificent luminaries: Rigel, positioned at his left foot, and Betelgeuse, which marks his right shoulder. Though both are supergiants that dwarf the Sun, the pair displays a contrast immediately evident to the eye: Rigel (Beta Orionis), listed among the brightest stars, is blue-white, while Betelgeuse (the Alpha star) is reddish. Rigel is some 11,000 K at its surface, whereas Betelgeuse, a prime example of a cool "red supergiant," is but 3700 K or even cooler. The common links between the blue and red supergiants are their dimensions and their similar intrinsic luminosities, which are typically 10 to 15 magnitudes – ten thousand to one million times – brighter than our Sun. But though Rigel is large, Betelgeuse, the cooler of the two, must be much the larger in order to be as bright. Though different, they – and the members of the two classes in general – are connected closely by the forces of stellar evolution, which can cause the blue and red supergiants to transform themselves into each other. Examination of the red supergiants also further outlines the distinctive natures of the luminous blue variables, which by contrast with the

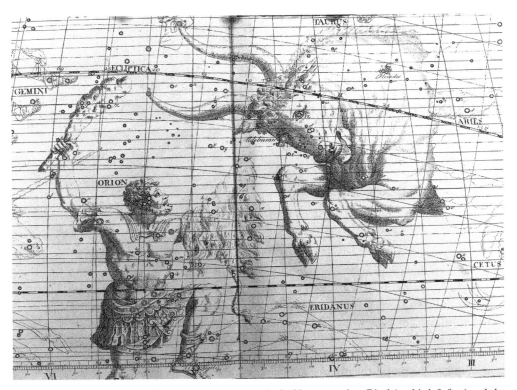

Figure 6.1. Magnificent Orion contains both the blue supergiant Rigel (on his left foot) and the archetype of red supergiants, Betelgeuse (on his right shoulder). [From Flamsteed's *Atlas Coelestis*, 1781 edition, courtesy of the Rare Book and Special Collections Library, University of Illinois at Urbana–Champaign.]

"red variety" of such variables (astronomical nomenclature notoriously inconsistent), are seen to be a distinctive class unto themselves.

In spite of their seeming number, the red supergiants are remarkably rare, more so even than the O stars. There is perhaps only one for every million or so stars in the Galaxy, and maybe far fewer; only about 200 are catalogued and studied. Nevertheless, they are so intrinsically luminous that several appear to the naked eye. Two are even of the first magnitude, and in a sense define northern winter and summer seasons. As northern weather warms to spring, and Orion and Betelgeuse set, Scorpius rises. At the scorpion's heart lies the other bright red supergiant, Antares, whose very name evokes its color: "ant-Ares," meaning "like Mars." How ironic that these two principal examples belong to antagonistic constellations that are forever coupled, as Scorpius was placed by the gods in the sky so that the fallen Orion need not look upon his slayer.

A list of naked-eye red supergiants is given in the accompanying table. Oddly there are no red supergiants of the second apparent magnitude, and we must drop to third before Alpha Herculis is encountered. This prominent and famous star is

124

Naked-eye red supergiants

| Star | Apparent visual mag. (V) | Spectral class[a] | Association | Distance (ly) | Absolute visual mag. ($M_V$) | Radius (AU) | Variable class[b] |
|---|---|---|---|---|---|---|---|
| Betelgeuse | 0.50 | M2 Iab | Ori OB1 | 430 | −5.1 | 3.6 | 5.8 yr + irregular |
| Antares | 0.96 | M1.5 Ib +B2.5 Ve[c] | Sco OB2 | 600 | −5.4 | 4.2 | Lc? |
| Alpha Herculis | 3.48 | M5 Ib–II | ... | 400 | −1.9 | 2.0 | SRc |
| Eta Persei | 3.76 | M3 Ib–IIa | ... | 1300 | −4 | ... | ... |
| Mu Cephei | 4.08 | M2 Iae | Cep OB2 | 2000 | −7.3 | 5.7 | SRc–Lc |
| 119 = CE Tauri | 4.38 | M2 Iab | ... | 2000 | −6 | 2.9 | SRc |
| Psi-1 Aurigae | 4.91 | M0 Iab | ... | ... | −6 | ... | Lc |
| VV Cephei | 4.91 | M2 Iaep +O8 Ve | Cep OB2 | 2000 | −8 | 8.8 | Lc (small); eclipsing, period, 20.2 yr |
| KQ Puppis | 4.97 | M2 Iabpe +B2 V | ... | 3000? | −6 | ... | Lc? |
| HR 8164 (Cephei)[d] | 5.66 | M1 Ibep +B2pe + B3 V | Cep OB2 | 2700 | −5 | ... | ... |
| 6 = BU Geminorum | 6.39 | M2 Iab | Gem OB1 | 4900 | −6 | ... | Lc; eclipsing, period 32 yr? |

*Notes:*

[a] e and p, respectively, stand for "with emission" and "peculiar"

[b] SR: semi-regular; L: irregular; c is an old designation for a "supergiant"

[c] visual companion, 3 seconds of arc from Antares

[d] HR means "Harvard Revised," the designation of the *Bright Star Catalogue*

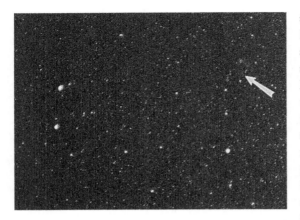

Figure 6.2. The star 6 Geminorum, its position indicated within its prominent constellation, is the faintest of the naked-eye red supergiants. Castor and Pollux lie to the left; the fuzzy ball just above the arrow is the open cluster M35. [Author's photograph.]

also called Rasalgethi, meaning "the Kneeler's Head," referring to the constellation's more-ancient name. There are two others of fourth magnitude, Eta Persei and Mu Cephei. Eta Persei lies at the northern end of the stream of stars that makes the main body of the constellation. With Alpha Herculis it represents a transition between the bright giants and the supergiants. Mu Cephei is sometimes called "Herschel's garnet star" because of its reddish hue, which is deeper than its spectral type would imply, though the color is nowhere near as vivid as that of the carbon stars. Fainter than these, only the variable star VV Cephei is well known and important enough to be discussed in detail later. CE Tauri lies obscurely among stars of similar apparent brightness in western Taurus, though Psi-1 Aurigae, located near a prominent stream of stars in southern Auriga, is a little easier to find. Both, along with the southern hemisphere's KQ Puppis, are erratic variables. Of the eleven M supergiants listed, three – Mu and VV Cephei plus HR 8164 – are grouped in one constellation, testimony to an intimate connection among such stars. Finally, at the bottom of the list lies 6 Geminorum, a challenge to the unaided eye. VV and Mu Cephei, the most intrinsically luminous of the set, represent the red supergiants in the HR diagram of the previous chapter.

## Spectra

Like red giants, red supergiants are all spectral class M, the coolest of the classic classes, with temperatures below about 3800 K. There is some overlap in luminosity between the very brightest of the asymptotic giant-branch stars and the dimmer red supergiants, the two mixing together around absolute visual magnitude −2 to −4. The terms "giants" and "supergiants" refer to more than just the luminosities and sizes of the stars. In a deeper sense, they refer to evolutionary condition. The supergiants evolve from the upper main sequence with original masses 10 or more times that of the Sun, and their internal structures can be very different from the stars on the AGB, which start at lower mass. This difference is reflected in the extreme spectral classes and temperatures of the two groups. For the most part, red supergiants are largely confined to warm M: of the naked-eye stars, only Alpha Herculis is as cool as class M5. The truly "cold" evolved M stars – M5 to M9 or so – are all asymptotic branch giants.

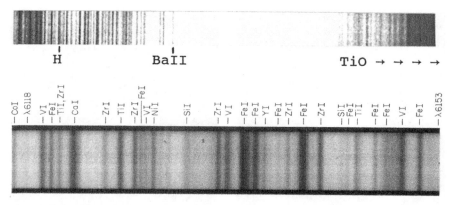

Figure 6.3. The titanium oxide bands of Mu Cephei (*top*) are relatively weaker than are those of the giants. Most of the lines in this blue-green spectrum (presented from 4300 Å to 5200 Å) are made by metals. An iron line is blended with the indicated hydrogen. A detail (*bottom*) of the red spectrum of Betelgeuse only 35 ångstroms wide shows about one observed line per ångstrom unit. With the exception of silicon, all the lines are of neutral metals. [*Top*: From *An Atlas of Representative Stellar Spectra*, by Y. Yamashita, K. Nairiai, and Y. Noromoto, University of Tokyo Press, 1978; *bottom*: The Observatories of the Carnegie Institution of Washington, courtesy of R. and R. Griffin.]

At the somewhat higher temperatures of the red supergiants, the titanium oxide bands that are so characteristic of the M stars are weak. Yet the spectra of these stars, while lacking the deep molecular absorptions of their cooler cousins, are nevertheless quite rich, showing vast numbers of lines created by neutral atoms, mostly metals. Class A supergiants can be distinguished from giants and dwarfs by the widths of their hydrogen lines. Class M supergiants can be separated from lesser stars of the same temperature class by their much weaker neutral calcium lines, weaker titanium oxide bands, and stronger hydrogen lines (which are just barely visible).

From such data, we can locate the supergiants and place them in context within the plane of the Galaxy and demonstrate their memberships within OB associations and a few young clusters, as done for giant stars. Even among the naked-eye supergiants, the relation is evident, as three in the accompanying table belong to the Cepheus OB2 association. The intimate relation between the red supergiants and the main sequence stars of class O, as well as their similar luminosities, clearly demonstrates that one kind of star (the O dwarf) evolves into the other (the supergiant).

The spectra of three of the stars in the naked-eye list are composites, showing the characteristics of more than one class: in these instances a B- or O-type spectrum is superimposed upon the background from the M supergiant (lesser main sequence stars may well be present, but would be invisible in the bright star's glare). These are close doubles that are inseparable optically, although orbital motion can be detected by Doppler shifts of the absorption lines. HR 8164 even has *two* B-type companions. Antares might be included as a fourth example, although the blue fifth magnitude neighbor can be seen through a small telescope about 3 seconds of arc from the magnificent red supergiant. Similarly, Alpha Herculis has a companion 5

Figure 6.4. (See also Plate XV.) An image of Betelgeuse (upper left in Orion) taken with the Hubble Space Telescope shows the star's disk and a bubble of hot gas on its surface. [A Dupree (Harvard-Smithsonian Center for Astrophysics), R. Gilliland (STScI), NASA, and ESA.]

seconds of arc away. Whatever the current mass of the red supergiant, it had at one time to be more massive than the companion dwarf, which it has now left behind on the main sequence. That most of the known companions are fairly massive B stars implies that the initial mass of the current M star had to be really large, in the realm of the O stars, again connecting supergiants with the upper main sequence.

## Size

The dimensions of the supergiants are truly astounding. In spite of their great distances, several are large enough to have had their angular diameters measured by various kinds of interferometry, including Betelgeuse, Antares, Alpha Herculis, and Mu Cephei. At a distance of 430 light-years, Betelgeuse is angularly one of the largest stars in the sky, some 0.055 seconds of arc across. It is exceeded only by a few relatively nearby giants like R Doradus (0.057 seconds of arc), R Leonis, and W Hydrae. From the angular diameter and estimated distance, we find that the radius of Betelgeuse is about 700 times that of our Sun, or 3.6 astronomical units. If placed in our Solar System, it would extend past the main asteroid belt to half-way between the orbits of Mars and Jupiter. Antares has an angular diameter only somewhat smaller, 0.046 seconds of arc. From its distance of about 600 light-years, however, it is physically larger, 80% the orbital radius of Jupiter. Alpha Herculis, with an angular diameter around 0.034 seconds and a distance of 400 light-years, is, with a radius of 2 astronomical units, notably smaller, though still larger than most giants (consistent with its being a crossover kind of star). Betelgeuse is so big that its apparent disk has actually been imaged by the Hubble space Telescope.

Some stars vault over high technology and provide the means for their own measurement. If two stars orbit each other very closely, there is a reasonable chance

128

that the orbit will be oriented such that from Earth each star will cut in front of the other to produce an eclipse twice each orbital period. The most-famed example is Algol, Beta Persei, which dims by a full magnitude every 2.9 days as a small bright B star partially hides behind a fainter K giant. In between two such primary eclipses is a much smaller dimming as the B star cuts out part of the light of its bigger companion.

There are several of these "eclipsing binaries" found among the supergiants, among the most famed, VV Cephei. The primary eclipse, produced when the brilliant O dwarf companion passes behind the M2 supergiant (which to the eye dominates the system), takes place every 20 years, when the combined light of the two stars decreases by about half a magnitude. The dramatic difference in size between most giants and the supergiants can be seen in the duration of the eclipse. The Algol eclipse lasts but a few hours. (It is in fact a partial eclipse, but even if it were total it would last but a day). But the O8 star of VV Cephei disappears for an amazing 1.2 years. From the "light curve," the graph of magnitude plotted against time, we can determine the tilt of the orbit, and from the Doppler shifts in the spectrum lines we can find the orbital speed. The duration of the eclipse then yields an accurate measurement of the size of the supergiant. The M supergiant has a radius 1900 times that of the Sun, or 8.8 AU, just smaller than the orbit of Saturn!

Mu Cephei is roughly comparable. From the star's parent OB association, astronomers estimate a distance of 2000 light-years. Its measured angular diameter

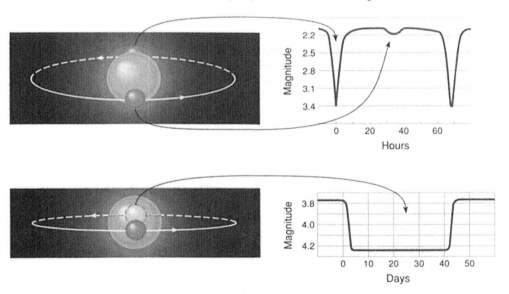

Figure 6.5. Stellar eclipses occur when one component of a double star cuts in front of the other. Graphs of brightness (in terms of magnitude) plotted against time are seen for each system. (*Top*) A partial eclipse and (*bottom*) a total eclipse of a small bright star by a large star. [From *Astronomy! A Brief Edition*, J. B. Kaler, © 1997, used by permission of Addison Wesley Educational Publishers Inc.]

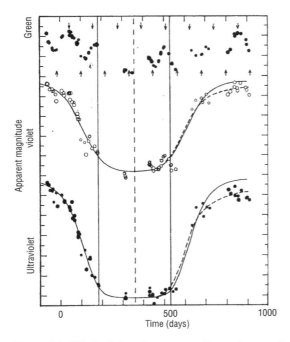

Figure 6.6. VV Cephei undergoes an eclipse when an O dwarf hides behind a red supergiant, as shown by graphs of apparent magnitude versus time. From top to bottom we see the eclipse in green, violet, and ultraviolet light. The marks at left for the top two curves are 0.1 magnitude apart, for the bottom, 0.2 magnitudes. The eclipse is particularly obvious at shorter wavelengths, since these are dominated by the hotter star. At longer wavelengths the M supergiant dominates and the eclipse is only barely visible; there we see the erratic variations of the M star. The solid vertical lines in the graph represent the actual disappearance and reappearance of the O star. The eclipse is not sudden, in part because some of the light is caused by matter flowing between the stars. [From an article in the *Publications of the Astronomical Society of Japan*, 1980, by M. Saito, H. Sato, K. Saigo and T. Hayasaka.]

of 0.0177 seconds of arc yields a physical radius of 5.7 AU, a bit larger than Jupiter's orbit. Such a measurement, however, is interpreted by assuming the star to be a uniform disk, whereas (as seen for the Sun in Chapter 1) stars are really darker at their edges than at their centers. If we include this "limb darkening," Mu Cephei must be larger, but by a degree difficult to know because of the vague edge possessed by such a huge, distended body. Another estimate of dimension is available from luminosity and temperature. A temperature of 3700 K, found from the spectrum, suggests that we should decrease the absolute magnitude (that is, brighten the star) from −7.3 to −8.8 to allow for radiation in the invisible infrared, where the star emits furiously. The temperature and resulting bolometric luminosity (a quarter-million times greater than the Sun) then yield a radius of almost 6 AU, somewhat larger. Whatever the results, we are clearly dealing with an enormous star.

The immensity of stars like VV Cephei and Mu Cephei are difficult to comprehend. We take pride in sending spacecraft from Earth to Uranus, a journey that took nearly a decade, yet it could all nearly fit inside a single star. In volume, VV Cephei could contain nearly seven billion Suns.

## Variation and mass loss

The very sizes of these stars render them unstable, causing them to pulsate erratically. Of the 11 stars listed in the table of naked-eye supergiants, nine are noted as variable, and four carry classical variable star names. Alpha Herculis, Mu Cephei, and CE Tauri are semi-regulars whose brightnesses wander around with only a vague period. Alpha Herculis changes by about a magnitude over a rough 6-year interval, and also exhibits short-term fluctuations. Mu Cephei displays larger variations of a magnitude and a half over two or three years along with a 13-year oscillation. Five others are classified simply as "irregular" ("Lc," for "supergiant irregular"). In the blue part of the spectrum, VV Cephei is a regular eclipsing variable. But in the infrared, where the M-star component dominates the light of the hot O companion, the star behaves quite unpredictably, although a 116-day period has been suggested. KQ Puppis, ninth on the naked-eye list, exhibits somewhat similar behavior.

Betelgeuse is a special case, its proximity and brightness allowing great detail in observation. Variation is first implied by its Greek letter name. Johannes Bayer made it Orion's alpha star even though it is usually about 0.4 magnitudes fainter than the beta star, Rigel. Quite likely when Bayer assigned the names, Betelgeuse rivaled or even exceeded its blue-white constellation-mate. Modern examination in fact reveals that Betelgeuse displays irregular variations on the scales of days, weeks, and months, as well as a slow 6-year periodicity that can change its brightness by nearly a full magnitude.

The origins of such pulsations are only poorly understood, but they are almost certainly related to winds and mass loss. The gravity at an M supergiant's surface is but a ten-thousandth that of our Sun, a small fraction the value of Earth's. Given the enormous energy radiated by these stars, closing in on a million times solar, it is no surprise that they are losing mass and indeed have very complex circumstellar structures.

Alpha Herculis has a fifth magnitude companion at an angular separation of 5 seconds of arc, easily viewed in a small telescope. The secondary is itself a binary (unresolvable by eye, but detected spectroscopically) that consists of a G5 giant and an F2 dwarf. As these two orbit each other with a 52-day period, their spectrum lines shift back and forth as a result of constantly changing Doppler shifts. Some of

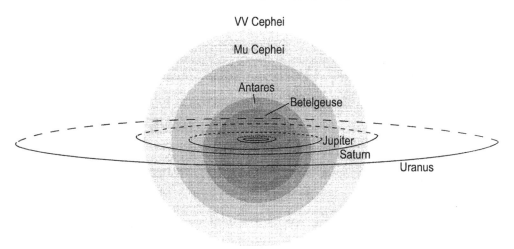

Figure 6.7. The largest supergiants, here represented by VV Cephei, Mu Cephei, Antares and Betelgeuse, would fill much of the Sun's planetary system.

131

the absorptions, however, do not partake in the shifts: they are stationary, and are also seen in the spectrum of Alpha Herculis itself. The supergiant, thought to be originally about 10 times the mass of the Sun, has lost so much matter that the expanding cloud has enveloped the companion, which is at least 550 astronomical units away from the supergiant, 275 times the supergiant's radius and 14 times the radius of our whole planetary system. The star appears to be losing mass at a rate of about a ten-millionth of a solar mass per year, hardly a record, but substantial enough.

A comparable cloud surrounds Antares, its spectrum lines also superposed upon the B companion. However in this case, the class B secondary star is hot enough to ionize the stellar ejecta, and consequently we see a small nebula around the red supergiant. Because Antares is both larger and more luminous than Alpha Herculis, its mass loss rate is some 50 times greater. Anomalous lines are also seen in VV Cephei's spectrum that show gas streaming from the great supergiant toward, and around, its companion.

The best observed of all these red supergiants, however, is still Betelgeuse. It is close enough to have been imaged, first through interferometry and then directly by the Hubble Space Telescope. We see what appears to be a hot spot, perhaps some kind of convective bubble or a feature related to atmospheric turbulence. Although we do not understand the origin of this structure, we begin to see the start of a real revolution in astronomy in which we can probe the disk of another star.

We know more about the extent and structure of Betelgeuse's enormous and complex circumstellar system. Interferometer data reveal a partial ring of dust with a radius about three times that of the star. The dust actually appears to be inside the chromosphere of the star, a cool but active external layer described in Chapter 2. (In the Sun, the chromosphere is a thin layer of cool red gas separating the bright photosphere from the extended corona). Observations made with an infrared interferometer show a ring of dust that peaks about 50 stellar radii away. In addition, more dust has been detected by directly observing blue light scattering from dust grains up to 90 seconds of arc from the star, 12,000 AU, 3,000 times the stellar radius. If your eyes were sufficiently sensitive you could actually see this great system as a disk with no optical aid at all. Where there is dust – probably silicates of some sort – there must be gas, and it too is found out to 1000 or so times the stellar radius.

Interferometer data also suggest the possibility that Betelgeuse has companions, one 20 or so stellar radii away and another perhaps at only 2 radii. The inner star may orbit the supergiant in only two years, and if it exists may actually be inside Betelgeuse's chromosphere. It may be responsible for producing the distorted half-moon shape of the inner dust cloud. On the other hand, the hot spots now detected on the surface of the star must have a profound effect on the dust's visibility. They are also probably connected with the heating of the chromosphere and illustrate the enormous turbulence of the whole system as well as its great internal instability. We have only begun to explore this remarkable system, its easy accessibility leading us to a fuller understanding of supergiants in general.

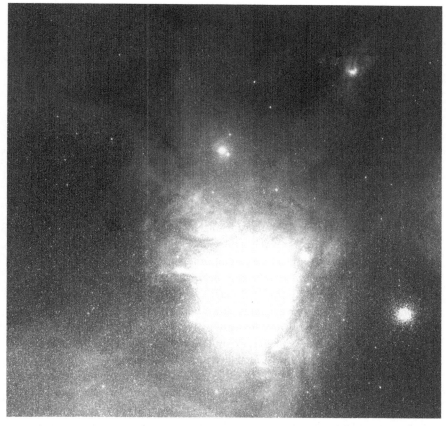

Figure 6.8. A small diffuse nebula surrounds cool Antares. Unlike real diffuse nebulae, much of the cloud is from mass lost by the supergiant and ionized by the hot B-star companion. The globular cluster M4 is seen to the right of the star; Rho Ophiuchi (at the western end of its dark cloud) is up and to the right. [© National Geographic–Palomar Observatory Sky Survey, reproduced by permission of the California Institute of Technology.]

## Early evolution

We can connect the largest stars in the Universe (the cool red supergiants) with the brightest stars (the blue supergiants), and both of these types with their lesser cousins, the red giants, through the theory of stellar evolution. We cannot watch individual stars go through the whole process, as their lifetimes are vastly too long. Instead we tie the individual parts together with theory, which attempts to fit what we see.

Though theoreticians may yet place particular kinds of stars in the wrong sequence, the overall picture is clear. Lower-mass stars, those from somewhat less than a solar mass to about 10 solar masses, turn into giants and then into the cool, windy asymptotic branch giants that precede the planetary nebulae. Their fate is to die as white dwarfs. Above 10 solar masses, however, the evolutionary tracks on the HR diagram lead into the realms of the blue and red supergiants. But the difference between the giants and supergiants is much greater than size and luminosity. Different interior developments do not allow the supergiants to become white dwarfs; instead, they make them explode. In between creation and explosion is a grand zoo of stars.

In the beginning, the evolution of high-mass stars is similar to stars farther down the main sequence except that high-mass stars do not live very long, theory in close agreement with observation (the reason an M supergiant can have a B dwarf companion). A 30 solar-mass star starts its life at spectral class O4, burning up the hydrogen in its core in only 7 million years. As it ages it brightens some, expands, and cools a bit to transform itself to type B1 at the right-hand edge of the main sequence. It may initially have ionized and lit its surroundings, enmeshing itself in a diffuse nebula much like the great Orion Nebula that lies on the sky between Betelgeuse and Rigel. The star reaches the end of its main sequence life at about the point where it can no longer produce a significant amount of ionizing ultraviolet, and the diffuse nebula turns into a "reflection nebula" in which interstellar dust grains mixed in with the gas merely scatter the star's blue radiation.

Finally, the nuclear furnace shuts down and the core rapidly contracts. The release of gravitational energy fires up fresh hydrogen in a shell around the core, and the star moves off the main sequence, expanding and cooling at its surface much like a giant except at a fairly constant luminosity. Since the star's total bolometric brightness does not change very much (at least in comparison with what happens to the lower-mass giants), cooler surface temperatures mean greater surface areas, so that upon leaving the main sequence the star, while brightening to the eye, also immediately becomes a large blue supergiant and then grows even larger into a red supergiant of class M. Once the high-mass star begins its death throes, it moves upon the HR diagram from B1 to K5 with remarkable speed. F, G, and K supergiants are therefore quite rare. As a star nears or enters the red supergiant state, the core becomes hot enough to fire its helium into a mixture of carbon and oxygen, which slows the pace of evolution, and the star stalls as a red supergiant. When we look at the distribution of stars on the HR diagram, we therefore see the red supergiants piling up in the upper right-hand corner, giving us a number of these awesomely huge stars even in the naked-eye sky.

The amount of mass lost during the metamorphosis from O main sequence to M supergiant is remarkable, although the exact figures still are not known. But what began life as a 30 solar-mass star may, by the time we study its red supergiant glow, be sliced in half or even more. The degree of mass loss is not theoretically predictable, and we must estimate it from crude measurements. But the winds powerfully influence the interior changes, so that we do not really know at any given external evolutionary state just what is happening inside. Where on the evolutionary tracks helium-burning actually begins depends critically upon both initial mass and the amount of matter sent back into interstellar space.

Wherever it occurs, the helium is used up in about a tenth of the time it takes to fuse the initial hydrogen. Here is the nexus, the point at which the supergiants separate from the giants. The giants can generally develop no farther, dying as degenerate carbon–oxygen white dwarfs. Only near the transition point on the main sequence, in the neighborhood of 10 or so initial solar masses, can the carbon and oxygen mixture in the giants fuse further to neon to produce a set of rare oxygen–neon white dwarfs. Betelgeuse and Rigel may in fact both fall into this realm, as their masses are at the lower end of the range that will make true supergiants.

A true supergiant is so massive, its core so large and hot, that it can enter progressively more advanced burning stages, the carbon and oxygen first fusing to become a mixture of oxygen, neon, and magnesium. When the carbon is gone, the oxygen–neon–magnesium core contracts, the core eventually heating to the point at which the dead mixture fires up to fuse into another mix of silicon and sulfur. At the same time, nuclear fusion continues moving outward into shells. Surrounding the

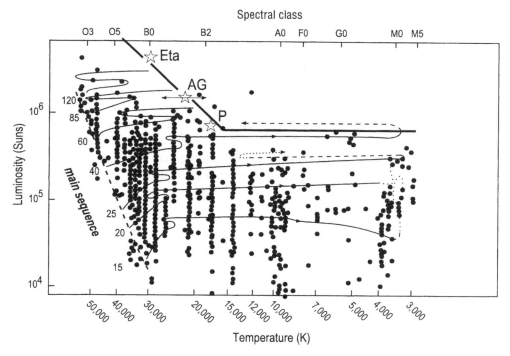

Figure 6.9. Supergiants of all kinds, as well as evolutionary tracks, are plotted according to temperature and luminosity. Initial masses, in solar units, are given along the main sequence. Once the hydrogen is consumed, the stars evolve horizontally as supergiants and develop rapidly into classes K and M. The dashed lines show where the stars are expected to be burning helium, dotted lines where they may be burning carbon. The heavy line is the Humphreys–Davidson limit, above which only a few stars stray, LBVs such as Eta Carinae, AG Carinae and P Cygni. [Adapted from the theoretical tracks of A. Maeder and G. Meynet and the observational diagrams of R. Humphreys and K. Davidson.]

neon (etc.)-burning core, we would – could we dive into the heart of the star – see a shell of carbon and oxygen fusing into oxygen, neon, and magnesium, which is in turn surrounded by another shell fusing helium into carbon and oxygen that is surrounded by yet another shell fusing hydrogen into helium, all the shells and the core contributing to supporting the star and creating its immense luminosity. Finally, the core "neon-burning" ends, and the silicon–sulfur core contracts, heats, and fuses to iron. Hosts of other fusion processes are going on as well in the surrounding shells. As some of the stars evolve through their various stages, they loop back across the HR diagram to become blue supergiants again, and may even re-loop to the red side. Exactly where on the tracks the stages take place, however, is uncertain. But how marvelous to realize the red supergiants were once as blue-white as the stars of Orion's belt. And may be again.

Iron is, in a nuclear sense, the most stable of elements. Energy can be produced by fissioning – splitting – elements heavier than iron or by fusing elements lighter than iron. Iron, however, can produce energy from neither process. To transform it into something else, you must add energy instead. One might think that the end would therefore be an iron white dwarf. One would be correct: but only for a fraction of a second before gravity, held at bay for the star's lifetime, becomes violently unleashed and the star implodes (and then explodes), the story to be told below.

Some of the red supergiants that we see may actually have advanced from helium-burning into one of these stages and may now be fusing their cores into neon or perhaps even into silicon or iron. The odds lie with helium-burning, as the other fusion stages take successively – and dramatically – shorter periods of time to go to completion (silicon-burning can run for only weeks). But there is no way of really knowing unless the iron core is complete and the star – whether as a red or blue supergiant – is seen to blossom brilliantly into the night sky.

## Hypergiants

Ascend the mass scale to the greatest and most brilliant of the supergiants, the hypergiants. Upon accounting for invisible ultraviolet and infrared radiation (that is, when using absolute bolometric magnitudes instead of visual magnitudes), we readily see that the brightest of the blue hypergiants are more luminous than the brightest of the red hypergiants. The upper limits to luminosity lie at or just below a rather well-defined upper bound called the "Humphreys–Davidson limit" that starts high in the O stars, slopes downward to about class A, and then levels off. Something odd is happening: the brightest of the blue hypergiants are not making it all the way across the HR diagram to become vast red stars like Mu and VV Cephei.

This limit – a strictly observational upper limit to luminosity – appears to be related to the theoretical "Eddington limit" at which the pressure of outflowing radiation on absorbing matter exceeds the inward pull of gravity, allowing for a dramatic increase in the rate of mass loss. The star effectively has zero surface gravity and simply cannot hold itself together. Instead it begins to tear itself apart.

Start with an array of high-mass main sequence stars from roughly 40 solar masses to about 120 solar masses. They begin their lives below the Humphreys–Davidson limit. Though they continuously lose mass through strong winds, they are more or less stable. When the stars exhaust their internal hydrogen, they move to the right on the HR diagram – toward lower temperature – to become supergiants and hypergiants. As they pass through the realm of the B stars, their diameters increase and so do their mass-loss rates. As stars over about 50 solar masses evolve, they encounter the sloping portion of the Humphreys–Davidson limit, really the Eddington limit, and because they are losing so much mass cannot evolve to class M, into the realm of the red supergiants. Instead, their evolution stalls. By this time they have lost maybe one-half their initial masses and they become the highly unstable LBVs whose winds blow at some hundred-thousandths of a solar mass a year. Some others enter a lesser "B[e]" class that consists of luminous B stars with emission lines. They seem to be related to the LBVs, and are surrounded by thick disks formed from equatorially ejected slow winds, while at the same time fast winds scream through the poles of the disks where they can escape, structures reminiscent of the eruption of Eta Carinae. Whether they are linked through evolution or are simply different versions of the same phenomenon is not known, nor is the role of binary companions.

For reasons not yet explained, an LBV will suddenly begin to lose mass at a rate much higher than normal, increasing the flow of its wind by a factor of 100 or more. In the case of massive Eta Carinae – or perhaps of a more advanced companion – the mass-loss rate increases by a factor of over 1000. The expanding gas cools and hides the star, presenting itself as a "false" stellar surface or atmosphere. However, the star inside keeps radiating energy at a constant rate, that is, its absolute bolometric magnitude stays roughly constant. Since the false atmosphere is cooling as it expands, the star seems to slide horizontally across the HR diagram, temporarily crossing the Humphreys–Davidson limit and radiating progressively more in the optical part of the spectrum at the expense of the ultraviolet. What we see as a visual outburst is therefore really a massive change in the bolometric correction (the difference between the visual and bolometric magnitude), which powerfully affects the visual magnitude. Once enough mass has been lost, some level of stability is restored, and the star returns to "normal," now perhaps with a small nebula around it. P Cygni, for example, is just sitting there waiting for another episode to begin. Stars like Cygnus OB2 #12 that are anomalously on the wrong side of the Humphreys–Davidson limit may be incipient LBVs.

Stars that begin their lives with masses below about 50 solar slide under the Humphreys–Davidson limit as they evolve, and can become the cooler hypergiants, even going so far as red class M. Rho Cassiopeiae and M33's variable A are among these. They are still close to the instability limit, and when they begin an episode of mass loss, their expanding false atmospheres make them appear cooler, and they radiate their energy in the infrared rather than in the visual. As do more normal supergiants in the 20 solar-mass range, these stars may turn around and evolve back to the hot side of the HR diagram, and they may eventually become the less-luminous LBVs.

Whatever form they take, the hypergiants eventually strip themselves of their entire original outer hydrogen envelopes, exposing the inner envelopes and cores that are rich in the by-products of millions of years of thermonuclear fusion. The LBV and other hypergiant activity seem to lead inexorably to the Wolf–Rayet stars, which are still losing mass at vigorous rates and have spectra rich in either carbon or nitrogen but are low on – or even completely devoid of – hydrogen. It is also possible that one kind of Wolf–Rayet star evolves into the other; if so, nitrogen-rich WN probably goes to carbon-rich WC. Alternatively, each kind may represent a separate pathway, the two possibilities revealing our ignorance of the details of high-mass stellar evolution.

## Mighty blasts

The final result of supergiant (and surely hypergiant) evolution is destruction as the star explodes as a supernova. The term is derived from "nova," meaning "new" in Latin, or in the context here, "new star." An ordinary nova involves a white dwarf in a double system, a subject to be explored under the smallest stars. A supernova, at least of the kind discussed here, is related to a nova only in the sense that it makes a grander show; the mechanism that produces the high-mass supernova is vastly different from the one that makes the nova.

We learned about the difference after Edwin Hubble was able to make measurements of the distances to galaxies. A "nova" had gone off in the middle of the nearby galaxy M31 in 1885, and when astronomers learned how far away the galaxy actually was, they could appreciate the immensity of the blast and also realize that earlier "new stars," those examined by Tycho Brahe in 1572 and by Johannes Kepler in 1604, were of the same broad variety. Examination of old Chinese writings eventually disclosed that an average of about one supernova is observed in our own Galaxy every 200 years or so.

Supernovae are so rare that (except for reconstruction of Tycho's and Kepler's observations) they have been studied only in other galaxies; none has been seen here since Kepler's star, although Supernova 1987A in our nearby companion, the Large Magellanic Cloud (at the distance of only 150,000 light-years), came close. Observation early in the twentieth

Figure 6.10. (See also Plate XVI.) Supernova 1987A, the aftermath of the evolution of a blue supergiant, exploded in the Large Magellanic Cloud. The supernova is the bright "star" toward the lower left. The bright, diffuse object toward upper right is the Tarantula Nebula. [AURA/NOAO/NSF.]

century revealed two very different kinds of supernova, those with hydrogen emission lines in their spectra (Type I) and those without (Type II). Later studies of the Andromeda galaxy also revealed two generally different kinds of stars, ancient ones with few heavy atoms residing in a spiral galaxy's bulge and halo (called Population II), and younger ones with a chemical composition more like that of the Sun that reside in the Galaxy's disk (Population I). To everyone's initial confusion, Type II supernovae were found exclusively in galactic disks of Population I, whereas Type I were found everywhere, including in Population II galactic halos.

The Galactic disk contains the raw material out of which stars are formed, and since massive stars live only short periods of time, they must all be within its confines. That Type II supernovae were found only in galactic disks therefore clearly indicated that they are related to massive stars. The developing theory of stellar structure and evolution allowed an explanation. These explosions follow the creation of an iron core within an evolving supergiant. Upon its completion, the core survives for a brief moment as an iron white dwarf, and then, upon discovering that it has no real support and nothing it can fuse into, the core collapses at ever-increasing speed, in a fraction of a second falling from the size of the Earth to the size of a small town at a velocity approaching that of light. In the collapse, the iron is broken down into its component atomic constituents, protons, neutrons and electrons, the protons and electrons merging to become more neutrons as the core shrinks to a tiny ultradense "neutron star." It may indeed collapse even further, into a black hole, both these endpoints subjects for the next chapter.

At densities approaching that of nuclear matter ($10^{14}$ grams per cubic centimeter, comparable to the Earth stuffed into a ball only 200 meters across), the core bounces back and a powerful shock wave – similar to a supersonic jet aircraft's sonic boom – tries unsuccessfully to move outward through the surrounding massive stellar envelope, the part that surrounded the once-iron core. The shock, however, has help. The merger that creates neutrons from protons and electrons also creates vast numbers of tiny nearly-massless (maybe really massless) particles called neutrinos. These curious specks of energy move at, or nearly at, the speed of light, but hardly react with anything at all. Those made in the center of the Sun in the proton–proton chain fly out unimpeded and can pass through the Earth. But the density is so high surrounding the collapsed core of the supergiant that the neutrinos actually push on the gas being hit by the shock wave, releasing it to roar outward through the star. A few hours later, when the blast reaches the surface, we see the star light up in detonation, whence it reaches a typical absolute visual magnitude of $-17$. If Betelgeuse were suddenly to go off in this way it would shine in our sky almost as brightly as the full Moon, and we could read by its light. The supernova of 1006, recorded by the Chinese, came close.

Proof of such processes came from a pair of neutrino detectors placed deep within the Earth for the purpose of monitoring solar neutrinos. The great 1987 supernova that erupted in the Large Magellanic Cloud used to be a blue (Type B1 Ia) supergiant that began life at about 20 solar masses (it exists no more). A few hours before anyone saw the star actually explode, a handful of neutrinos were absorbed and counted by the detectors, just the amount predicted by theory for such a distant supernova. The theory works. Supernovae, at least those with hydrogen in their spectra, those from massive stars, result from the collapse of iron cores that develop within supergiants. At the peak of such a supernova, the rate of released energy (most of which is in neutrinos) exceeds that of all the stars in the visible Universe.

Some of the hydrogen-less Type I supernovae may also be such beasts. Type I explosions break down into two subtypes (Ia and Ib) that are respectively discriminated by the presence or absence of silicon. The element is absent in Type Ib supernovae, which are also the fainter of the two kinds and, like Type II events, occur in galactic disks undergoing active star formation. It is Type Ia that is seen in the galactic halos. The Ib variety is reasonably interpreted as supernova explosions that involve Wolf–Rayet stars from which the hydrogen has already been removed by fierce winds.

In the explosion, the by-products of millions of years of nuclear fusion – helium, carbon, oxygen, and more – are blasted into space. Moreover, the expanding shock wave within the supernova is at such a high temperature (in excess of 100 billion kelvin) that fierce explosive burning takes place within the oxygen and silicon shells that creates additional new elements, including a good fraction of a solar mass of radioactive nickel. It is nickel's decay through cobalt into iron that is largely responsible for a supernova's visual light, the iron thence blasted into space.

Within giant stars neutrons can be captured slowly (the "s-process") to produce elements up to bismuth. Within the maelstrom of a supernova (the exact site is still quite uncertain) neutrons can be captured like machine gun bullets, allowing the instantaneous creation of very heavy isotopes that will decay into the heaviest of know elements, uranium, and even into plutonium and quite likely beyond. This "rapid neutron capture" process (the "r-process") can also make other elements heavier than iron. Many of the more exotic elements like gold and silver – including all that on Earth – seem to be made almost exclusively by the r-process.

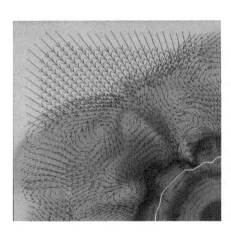

Figure 6.11. (See also Plate XVII.) A computer simulation shows a quarter view of the turbulent core of a supernova only 120 kilometers across a few-hundredths of a second after the collapse. The wavy white line is the wall of outwardly pushing neutrinos. A neutron star is being created within. [A. Burroughs and J. Hayes.]

Why some stars blow up as red supergiants (at least so we think, though we do not really know), some as blue supergiants (like the one in the Large Magellanic Cloud), and others not until they reach Wolf–Rayet status, is still a

Figure 6.12. The Crab Nebula is the gaseous remnant of the great supernova of 1054. [Palomar Observatory, California Institute of Technology.]

mystery that is related to uncertain mass-loss rates and to the locations in the HR diagram at which various nuclear-burning stages take place. The best candidates for such explosions are Wolf–Rayet stars like our local Gamma-2 Velorum, those at the centers of ring nebulae, certainly LBVs like Eta Carinae, and perhaps even Betelgeuse or Rigel. One of them could go tonight.

## Markers

The destruction of such massive stars leaves different kinds of remains. First, we see an ejected envelope, the part of the star blasted into space, as an expanding gaseous remnant. The best example is the Crab Nebula, positioned at the end of the southern horn of the constellation Taurus. The object, easily visible in a small telescope, is at the position of a glorious supernova recorded in the year 1054 by the Chinese, one that shone with the apparent brightness of the planet Venus though over 6000 light-years away.

The Crab nebula, expanding at a speed of 1300 kilometers per second, is enormously energetic. It is also one of the brightest sources of radio radiation in the sky. The spectrum of the radiation shows that it – like much of its optical radiation – is not produced because it is a hot body, but by electrons spiraling in a magnetic field at speeds near that of light. In deference to atomic particle accelerators on Earth, this distinctive natural version is called synchrotron radiation.

About 100 such "supernova remnants" are observed in the Galaxy, most by

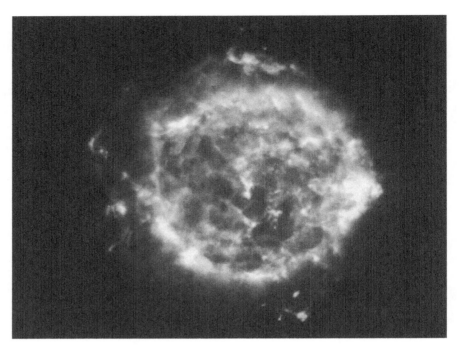

Figure 6.13. Cassiopeia A, a shock-wave remnant from a supernova that exploded about 1650 behind a cloud of interstellar dust that rendered it optically invisible, glows brightly in the radio spectrum as a result of its synchrotron radiation. [VLA/NRAO/AUI.]

observation of radio waves that can punch through the thick dust that pervades the Galaxy's disk where the majority of these remnants are found. The presence of synchrotron radiation gives the remnants away, distinguishing them from the far more common diffuse nebulae. Supernova remnants must eventually dissipate into the interstellar gloom, carrying with them the elemental richness created within the heat of the parent supernovae. Much of the heavy stuff of which the Earth is made, including all the iron, seems to have come from them. And long after the brightness of the expanding gas fades away, they still make themselves known by the expanding blast waves that heat and light the ambient gas of the interstellar space.

The force of the shocks is so great that they can apparently turn grains of interstellar carbon into tiny diamonds whose spectral signatures are found within the dark clouds of interstellar space. The diamonds themselves, so small that a microscope is needed to see them, are found in meteorites, helping to link the formation of the Sun and of ourselves to our origins in interstellar space. More significantly, the force compresses clouds of interstellar matter, initiating the collapse of the thin gas into new stars, as may have happened to make the Sun and its family.

Supernovae leave more lasting remains in the form of "cosmic rays," a misnomer that implies electromagnetic radiation but instead consists of enormously high-speed particles that pervade interstellar space. Cosmic rays were first discovered as a result of "showers" of subatomic particles that spray in a narrow cone onto the Earth. High-flying balloons and then spacecraft demonstrated that the showers were being caused by ordinary atomic nuclei that had been accelerated to extraordinary velocities, to nearly the speed of light. Most were simply hydrogen ions – protons – but other nuclei are found up to and even well beyond iron. At their highest energies, a single nucleus can have the energy of a professionally served tennis ball! Cosmic rays plague astronomical equipment, especially in space, where they activate detectors and produce false "stars." They have even been known to wake astronauts by producing retinal flashes of what seem to be light. They are firmly believed to be launched by supernovae, accelerated by the expanding remnant, and confined by the magnetic field of our Galaxy (produced by the Galaxy's rotation). A large fraction of the energy of the Galaxy is locked up in them, and they play a powerful role in the star-formation process (to be examined later).

Since massive stars tend to belong to groups, supernovae and their remnants can clump together and consolidate their power, resulting in huge bubbles blasted in the interstellar medium that can be hundreds of light-years across. New massive stars forming from condensation of matter caught in the expanding waves then generate new supernovae to produce bubbles within bubbles in an outwardly expanding chain of alternate destruction and creation. The shocks apparently even have sufficient strength to burst through the disk of the Galaxy, sending thin hot plumes of gas far into the halo, from which it rains back to the Galactic plane.

At the center of a supernova remnant should lie the collapsed remains of the imploded core, a tiny ball of neutrons. The neutron star is but one of four possibilities for the conclusion of stellar evolution, the most common, the white dwarf, the remains of an ordinary giant star. In the next chapter we will pursue both white dwarfs and the neutron stars, as well as the two other endpoints: the black hole and nothing at all – complete annihilation – which may be related to

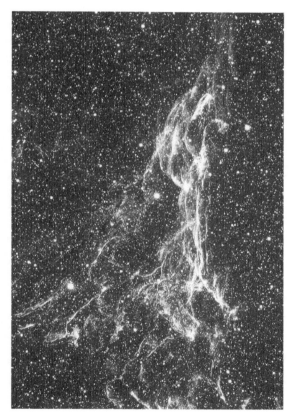

Figure 6.14. The Veil Nebula in Cygnus is a portion of a giant blast wave that has swept up and heated interstellar matter. It was produced by a supernova that exploded perhaps 50,000 years ago. [AURA/NOAO/NSF.]

the heretofore ignored Type Ia supernovae. The Type Ia supernovae are, in turn, almost certainly linked back to white dwarfs, completing something of a cycle, and allowing a self-consistent discussion of all the end possibilities together under the title "the smallest stars," to which we now turn.

# Chapter 7

# The smallest stars

Thickly scattered throughout the Galaxy is the debris of stellar evolution, the shadows of once-mighty stars. White dwarfs are the cores of suns in which nuclear fires once raged, and are destined to cool as dense celestial ashes until the end of time. They are comparable to our Earth in size, yet retain enough of the mass of their parent star to give them enormous density. Widely sprinkled among them are much rarer beasts, neutron stars, some behaving as pulsars, the products of the catastrophic evolution of massive stars. These are so strange as almost to defy description: great stellar remnants shrunk to the size of a small town containing the most dense matter known, sometimes spinning madly in space at rates approaching 1000 revolutions per second. Neutron stars are exceeded in their extreme state only by black holes, in which matter as we know it is crushed out of existence, their mass creating a gravitational field so strong not even light can escape.

Studies of stellar endpoints and of their binary interactions with more normal stars help reveal more about how stars work, and tell us as well how matter behaves under circumstances impossible to create in the laboratory, revealing more about the general nature of the physical world.

## White dwarfs

Stretched out in a long thin line on the HR diagram, 10 to 15 magnitudes below the main sequence, are the ultimate descendants of ordinary stars like the Sun, the white dwarfs. Once thought to be rare, they abound in the Galaxy, as this group is the final repository of the whole middle part of the main sequence, the same set of stars that produces the giants and their successors, the planetary nebulae.

The story of these stars goes back to 1844, when Friedrich Bessell discovered that neither Sirius nor Procyon moved through space in a straight-line path. He hypothesized that these two bright stars, respectively an A1 dwarf and an F5 dwarf–subgiant, and coincidentally the two luminaries of Orion's hunting dogs,

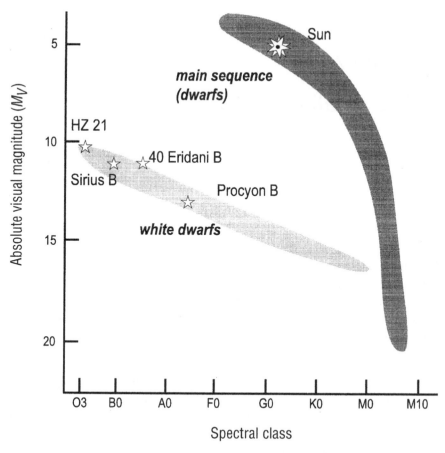

Figure 7.1. The lower half of the HR diagram shows a small selection of prominent white dwarfs stretched out in their dim string below the main sequence. They are placed here according to their temperatures relative to those of main sequence stars. They cool from left to right.

were being shifted by massive unseen orbiting bodies, or "dark stars." (The struggle to find the "dark matter" or "missing mass" of the Universe is nothing new.) The companion to Sirius was located about where expected just 18 years later by Alvan Clark upon testing the 18-inch lens of the refractor of Northwestern University's Dearborn Observatory. Sirius B is an eighth magnitude star. By itself it would be viewable with large binoculars, but it is rendered extremely difficult to observe by its proximity (less than 10 seconds of arc) to brilliant Sirius A, 10 magnitudes its superior. Procyon B, much fainter at 11th magnitude, finally fell in 1895 at the Lick Observatory in California.

The orbit of Sirius B relative to Sirius A, and the location of the center of mass, quickly showed that the small companion had a mass about one-half that of the bright star and comparable to that of the Sun. But no one knew what it was; the small star may simply have cooled and dimmed. The solution came from a similar star, the companion to the otherwise unremarkable 40 Eridani. This 9.6 magnitude star, 40 Eridani B, is easily seen in a small telescope, and itself has a companion, a faint 11th magnitude red dwarf of class M4 V featured in Chapter 2. These two orbit each other with a period of 250 years and are currently near maximum separation, about 9 seconds of arc. From the orbital motion, the white dwarf was found to have a mass nearly one-half that of the Sun, roughly comparable to the mass of Sirius B

But 40 Eridani B has a great advantage over Sirius B, as its primary star is both fainter and angularly farther away. As a result, astronomers could effectively observe its spectrum, not possible at the time for Sirius B. To everyone's astonishment, 40 Eridani B was an A star, one with strong hydrogen absorption lines. It therefore had to be hot. For the luminosity to be so low, the radiating surface area – and the radius – had to be very small. Thus 40 Eridani B, dramatically set off by itself low on Russell's original HR diagram, has the distinction of being known as the first "white dwarf," the name derived from the apparent color and the small size.

Shortly thereafter, in 1915, W. S. Adams of Mt Wilson Observatory discovered that Sirius B was also an A star. Ten magnitudes below Sirius A on the HR diagram, it must be only 1/100 the latter's size. Since Sirius A, an A1 star at 9000 K and absolute visual magnitude 1.4, has a diameter twice that of the Sun, Sirius B was estimated to have a radius only double that of Earth. From the binary orbit Sirius B was found to weigh in at about one solar mass (somewhat higher than typical), so it had to have an average density 100,000 times that of water, or 10,000 times that of lead. Modern observations place the temperature higher than originally thought, but yield conceptually the same results, a radius three-quarters that of Earth and a density over a million times that of water. These two stars, with Procyon B and two others, are listed in this chapter's table, which presents a variety of white-dwarf properties.

## White dwarf conditions

The crushing densities and pressures of white dwarfs are vividly revealed by their spectra. Like all stars, a white dwarf will have its highest density at its center and its lowest in a thin atmosphere where the absorption-line spectrum is actually formed. The densities and temperatures in this outer region are nevertheless still extremely large. The atomic orbits from which an

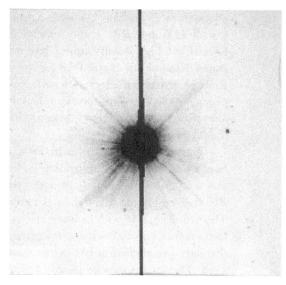

Figure 7.2. In this negative image, Procyon B is the tiny 11th magnitude dot to the right of the brilliant image made by naked-eye Procyon A. [R. Gilliland *et al.*, STScI, and NASA.]

Figure 7.3. The main photo shows 4th magnitude 40 Eridani A to the west of Orion, the inset a magnified view and the bright star's companion. The companion consists of the blended image of a white dwarf and a 1.5-magnitude-dimmer red dwarf. [*Main photo*: author; *inset*: © National Geographic–Palomar Observatory Sky Survey, reproduced by permission of the California Institute of Technology.]

absorption-line is formed can be disturbed, their energies shifted, by electrical effects from neighboring atoms. As a result, atoms are capable of absorbing photons that do not lie exactly at the "official" wavelength of an absorption line. A line formed in a high-density atmosphere will have a large number of such disturbed atoms. Its absorptions will therefore be wider than those formed in a low-density atmosphere where the atoms are well separated from each other. Thus class A supergiants, which because of their sizes have low densities, will have narrower hydrogen lines than class A giants, and these will have narrower lines than A dwarfs, allowing the determination of luminosity.

White dwarfs carry this behavior to an extraordinary extreme. The hydrogen lines are amazingly wide, providing evidence for almost unbelievable densities. But they are "unbelievable" only from our parochial perspective. In truth, the atom is almost all empty space, and there is a great deal of room for compression of matter. The hydrogen atom with its single electron is about a hundred-millionth of a centimeter across, but the nucleus – the proton – is 100,000 times smaller, which means that only a single part in $10^{15}$ (a thousand trillion) of the atomic volume is actually filled. The scale is seen by placing a grain of rice, representing the proton, in the center of a field of a sports stadium. The electron, which has no measured dimension, rides the outer seats.

The famed Cambridge astrophysicist Sir Arthur Eddington realized that for a

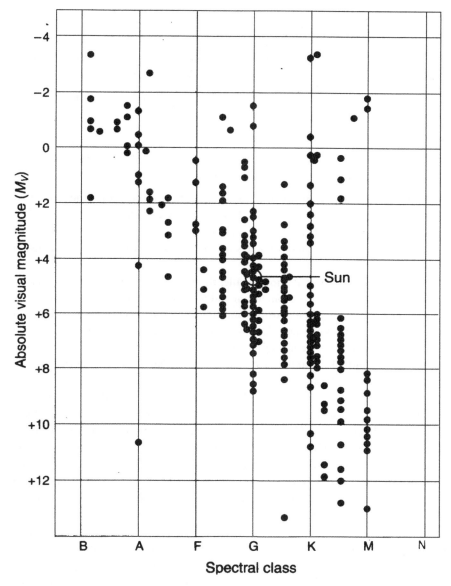

Figure 7.4. Henry Norris Russell's original HR diagram plots absolute visual magnitude against spectral class. One lonely star resides in class A near absolute magnitude 11, 40 Eridani B, the first recognized white dwarf. [From *Popular Astronomy*, 1914.]

white dwarf to exist, the gas had to be fully ionized (the electrons stripped from their nuclei), so that the atoms could be compacted together. But the stars still presented a conundrum. As they cooled, the atoms ought to recombine, the protons linking with the electrons, and the stars should swell; but they could not swell because of the intense gravities, some 10,000 times solar. The solution came from R. H. Fowler

| Star | Apparaent visual mag. (V) | Absolute visual mag. ($M_v$) | Diameter (Earths) | Class | Temp. (K) | Mass (Suns) |
|------|---------|---------|---------|-------|------|--------|
| Sirius B | 8.4 | 11.2 | 0.92 | DA | 27,000 | 1.03 |
| Procyon B | 10.9 | 13.2 | 1.05 | DA | 8700 | 0.62 |
| 40 Eridani B | 9.5 | 11.0 | 1.48 | DA | 14,000 | 0.44 |
| HZ 21 | 14.2 | 10.3 | . . . | DO | 50,000 | . . . |
| ZZ Ceti | 11.9 | . . . | . . . | DA | 12,000 | . . . |

(also of Cambridge) by an application of "quantum mechanics," the rules that govern the behavior of the very small. In an ordinary gas, atomic particles are free to take on any value of velocity or momentum (mass times velocity). But at extremely high densities they become subject to a law called the Pauli Exclusion Principle.

To define it, look first at another rule, the Heisenberg Uncertainty Principle, which states that a particle's momentum and its position cannot be simultaneously known. The uncertainty in its position multiplied by the uncertainty in its location roughly equals Planck's constant, $h$, a tiny number that relates the energy of a photon to its frequency ($E = h \times$ frequency). In units that relate centimeters, seconds, and grams, $h$ is around $10^{-27}$, so the principle's effect is significant only in the exceedingly small world of the atom.

Now define a six-dimensional "volume" that consists of the three directions of real space and three more of momentum whose axes are directed along the real coordinates of ordinary space. This strange six-sided volume is derived by multiplying all six "directions" together. The Pauli Exclusion Principle states that in a minimum six-sided volume of size $h^3$, no two particles with identical properties can exist; only one can be there at a given time. As a result, there can be at most two electrons within this minimum box, and they are required to spin in opposite directions. The practical result of the Pauli Exclusion Principle for free electrons is that any given small volume of space can contain only a certain number of particles moving at a particular momentum, or speed. We can add particles, but only at ever-increasing speeds. When the spatial volumes for a given momentum become filled, the gas becomes "degenerate," and the electrons produce an outward pressure. This "degeneracy pressure" is enough to hold a white dwarf up against gravity and forever keep it from collapsing.

In an ordinary gas, pressure is directly proportional to temperature and inversely proportional to volume (or density). Degeneracy changes the rules. In a degenerate gas, the pressure depends only on density and not on temperature at all. As a white dwarf and its degenerate gas cool, the electrons are packed at low velocities as tightly as possible and cannot slow down, and except near the stellar surface (which is not degenerate), recombination of electrons and protons into atoms cannot

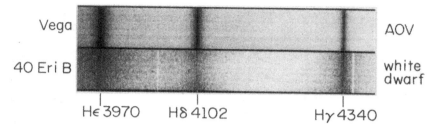

Figure 7.5. The hydrogen lines in the white-dwarf 40 Eridani B's spectrum (*bottom*) are much broader than those in the ordinary A dwarf Vega (*top*) as a result of the extreme gas pressure in the small, degenerate white-dwarf. The bright lines in the white dwarf spectrum are caused by street lights. [From *An Atlas of Representative Stellar Spectra*, by Y. Yamashita, K. Nariai, and Y. Norimoto, University of Tokyo Press, 1978.]

take place. The star maintains a constant internal pressure dependent upon its mass, and as time goes on cools with constant radius.

Most white dwarfs have masses somewhat over one-half that of the Sun, though a few, like Sirius B, approach or even surpass solar. A higher-mass white dwarf will be squeezed into a smaller space by its gravity, so that, at a given temperature, it will have a smaller surface areas and be less luminous than its lower-mass counterpart: just the opposite of main sequence behavior. This mass–radius effect begins on the descending tracks of the nuclei of planetary nebulae and continues on down to the faintest white dwarfs at the end of the sequence. It is quite obvious in comparing the masses and radii of the white dwarfs presented in the this chapter's table.

Under conditions of ordinary degeneracy there is no limit to the mass of a white dwarf. But as the mass grows, and the density becomes higher, conditions are no longer "ordinary." The velocities of the highest-speed electrons begin to approach that of light, and the theory of relativity must be taken into account. Using relativistic principles, Subrahmanyan Chandrasekhar showed in 1930 that there is a limit to the mass of a white dwarf, of a star supported by electron degeneracy. No white dwarf can be heavier than 1.4 solar masses, and, as expected, none is found. Beyond that mark the star must collapse to a smaller state, which might be a neutron star or a black hole. It might even explode as a result of violent contraction to become nothing at all. Chandrasekhar's profound discovery – it was later to earn him a Nobel Prize – was not accepted when he first presented it. Eddington simply rejected the concept of relativistic degeneracy out-of-hand, and his influence was so great that some two decades were to pass before the course of late stellar evolution could again be placed in the right direction.

The whole life story of a star can be told through its efforts to contract under gravity and the forces that keep it from doing so. In the early stages of a star's life, it is buoyed up by hydrogen burning. When the hydrogen runs out, gravity gets the upper hand and the core contracts to produce a giant star. Core contraction is temporarily halted again by helium burning and a second giant state (the asymptotic giant branch), but when the helium is gone, contraction resumes. In a lower-mass

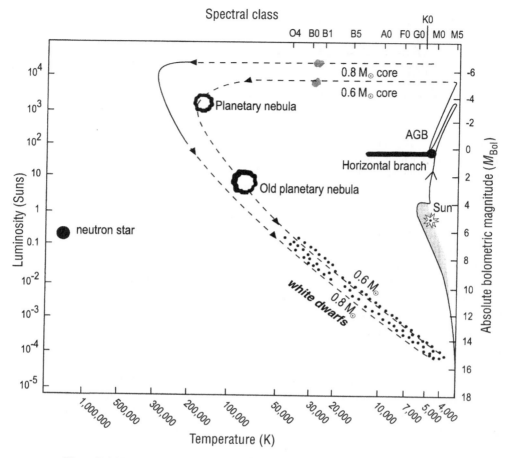

Figure 7.6. The whole flow of stellar evolution for lower-mass stars can now be shown, from the main sequence to the cooling white dwarfs. Evolutionary tracks are shown for white dwarfs of 0.6 and 0.8 solar masses (designated M.). The higher-mass stars on the main sequence (and in the giant state as well) are the brighter, but after nuclear burning shuts down, and the stars dim and cool, the relation is reversed. One neutron star lies near the left-hand edge. [Diagram based on work by I. Iben Jr.]

star, from about one to 10 solar masses, the core becomes degenerate, gravity loses the battle, and the star stabilizes forever as a white dwarf.

The size of a star's life-sustaining core, and the mass of its ultimate white dwarf, depend on the mass with which the star began its evolutionary journey. The lower the initial mass, the longer the star's life: below about 0.8 solar masses the lifetimes are so long that no such star has as yet been able to evolve off the main sequence. At the lower limit, the star will develop into a white dwarf a bit over one-half the mass of the Sun, the outer envelope lost through winds and the creation of a planetary nebula. The calculations show that final core mass climbs to maybe 0.8 solar masses for an AO dwarf with an initial mass of 2.8 solar, and to one solar mass for a B7 main sequence star (initially 4.5 times solar). Finally, somewhere around 10 solar masses – typically a B2 main sequence star – Chandrasekhar's limit of 1.4 solar masses is reached. As a result, white dwarfs arise from the main sequence between spectral classes G8 and B1. Above about 10 or 12 solar masses, stars do not become white dwarfs, the rare O stars evolving into supergiants, exploding as supernovae, and ultimately producing neutron stars and quite likely even black holes.

As the old nuclear-burning cores of giant stars, most white dwarfs are made from a mixture of carbon and oxygen. Only near the transition point, near the Chandrasekhar limit, will nuclear reactions go any farther, producing a small set of rare oxygen–neon white dwarfs. From this point on, white dwarfs cannot be freed from the interiors of the developing stars, the nuclear reactions continue, and the cores are fated only to implode, releasing the energy it takes to make a supernova.

## Weird atmospheres

White dwarfs present us with a wonderful and bewildering array of spectral and atmospheric properties. As more were discovered, they were found to fall along the entire spectroscopic sequence. Those like Sirius B, Procyon B, and 40 Eridani B, with powerful hydrogen lines, were classified as "DA," "D" for degenerate, "A" for their apparent similarity with the class A main sequence stars, with which they share the property of strong hydrogen lines. But there are other kinds. Some hot white dwarfs showed no hydrogen absorptions, but instead displayed those of neutral helium. Since neutral helium lines define the B stars, these were naïvely, though quite naturally, called "DB," while those with ionized helium absorptions were called "DO." Cooler white dwarfs, for which the Balmer lines are weaker or absent, were classified DF, DG, DK, and even DM on the basis of the strength of the ionized calcium lines (which are powerful in the Sun's spectrum), following the criteria used for the main sequence.

The parallel with the luminous stars on the main sequence and with giant and supergiant branches of the HR diagram is not appropriate, however. First, white-dwarf spectra embrace two other categories: stars that show only weak bands of molecular carbon, and some that exhibit no lines at all, called "DC" for "continuous." More important, and in complete contrast to normal stars, white-dwarf

Figure 7.7. White-dwarf spectra are arrayed according to decreasing temperature, from 50,000 K on the top to about 6000 K or so at bottom. The DO and DB stars, respectively, display ionized and neutral helium lines, but none of hydrogen, while the DA star gives us strong hydrogen but no helium. The DF and DG stars exhibit metal lines. [Palomar Observatory spectrograms, from an article by J. L. Greenstein, in *Stellar Atmospheres*, J. L. Greenstein, ed., University of Chicago Press, Chicago, 1960.]

153

spectra feature a curious exclusivity: the DA stars have *only* hydrogen lines, and the DB and DO stars have no hydrogen at all. Ordinary B and O stars, on the other hand, prominently feature the hydrogen Balmer lines.

The only explanation is physical separation of the atoms, rendering the surface gases of white dwarfs grossly different from those of ordinary dwarfs. The classifications, which had already been applied to the white dwarfs, then came to mean something quite different, no longer reflecting temperature but chemical composition. The term DA now means a white dwarf with a pure hydrogen atmosphere, and DB denotes degenerate stars with pure-helium atmospheric compositions in which the amount of hydrogen is less than a ten-billionth that of helium. The DO stars are just hot versions of type DB, but are still sometimes called DB.

We cannot recognize the DB stars below about 10,000 K, where the temperature is too low to produce helium lines (as for main sequence stars), although DA stars can be found to 5000 K or so before the hydrogen absorptions finally disappear. Some of the cooler classes could then be DB without our knowing it; at the very coolest end of the sequence, DA and DB can no longer be discriminated. For simplicity, the DB, DO, and other categories that do not show hydrogen lines are simply lumped together as "non-DA." The DA and DB (or more generally, non-DA) classes, far from indicating temperature, therefore run through the entire temperature sequence. Sirius B, the archetypal DA star, has an effective temperature of 27,000 K, which should qualify it for class B0.5, almost to class O. Even some planetary-nebula nuclei with temperatures above 100,000 K have been termed DA. Traditional spectral classification, which is based upon solar-type atmospheric composition, is therefore impossible. White dwarfs are placed on the HR diagram here according to their temperatures and the temperature scale of the main sequence. Though an improved, more complex, system is available (in which the above classes retain their old meanings and additional letters and numbers are added to reflect other properties and temperatures), the old "DA" – "non-DA" remains in common use.

The physical causes of the white-dwarf classes are fairly well understood. The surface gases of the DA stars are pure hydrogen because helium and heavier elements have settled downward to lower layers under the action of the powerful gravitational fields. The DB stars apparently have had their hydrogen stripped away or otherwise removed: they have almost none at all, and consist of carbon–oxygen cores surrounded by a helium layer that perhaps includes a mixture of carbon and metals. Their predecessors are probably various classes of planetary-nebula nuclei that have hydrogen-poor atmospheres, the separation beginning quite early after the stars leave the asymptotic giant branch.

Mysteries, though, still abound. Above 15,000 K, the non-DA variety numbers only about 15% of the total, whereas below that temperature, the fraction is roughly one-half. Even odder, there are no non-DA stars between 30,000 K and 45,000 K. Apparently as a white dwarf cools it can change from one type to another and even back again. How, we are not sure. Upward or downward diffusion of elements may be responsible, but other processes may be involved, none of which we understand very well.

## Variables

It seems surprising that stars as dense as these, with such enormous gravitational fields, could pulsate and vary in their brightnesses, but indeed they do. The HR diagram's instability strip through the F and G supergiants contains the Cepheid variables. These great stars vary by a few magnitudes and take days to perform one pulsation. The strip plunges downward across the main sequence at class A (where they vary mildly as "Delta Scuti stars") and below to the white dwarfs within the B temperature range, 10,500 to 13,000 K. Here the DA degenerates quietly chatter away, changing their brightnesses by 0.01 to 0.3 magnitudes, with periods of 3 to 20 minutes that befit their small sizes. The instability is caused by the same basic mechanism that makes a Cepheid pulsate. Here, a zone of hydrogen ionization beneath the stars' surfaces acts like a valve, alternately absorbing and releasing energy. However, the pulsations of these "ZZ Ceti stars," named after the prototype, are not as simple as those of the Cepheids. The Cepheids expand and contract as a unit, the whole star behaving in a concerted manner. The surfaces of the ZZ Ceti stars, on the other hand, expand and contract at the same time, some zones pushing outward while others pull inward.

Different classes of white dwarfs pulsate in different temperature regimes. Ascend the white-dwarf sequence to a point near 20,000 K, and find the region in which the DB stars pulsate in a little instability strip of their own caused by deep helium ionization. Far upward toward the realm of the planetary nebula central stars, near 130,000 K, are the rare "PG 1159 stars" (the name means "Palomar–Green Survey" $11^h$ $59^m$ right ascension, the coordinate position of the prototype). These too vary with periods of minutes, but owing to a region of *oxygen* ionization. A few actually are the nuclei of planetary nebulae, but most are not, showing either that these white dwarfs missed the planetary state or that the nebulae have dissipated. The planetaries in this category are all "O VI stars" (those with emissions of five-times-ionized oxygen, $O^{5+}$): a clue that has yet to be interpreted. The PG 1159s that are planetary nuclei also have longer periods for reasons that are unexplained.

## Explosions

About once a generation, we wander out on a dark night to admire the constellations and discover that some familiar figure is oddly distorted. There is a "new" bright star, a "nova," where none had been seen before. Several times over the twentieth century, brilliant first magnitude events have taken place: in Perseus in 1901, Aquila in 1918 (which set the century's record for apparent visual magnitude of $-1.1$), Pictor in 1925, Hercules in 1934, and Puppis in 1942; one in Cygnus in 1975 just

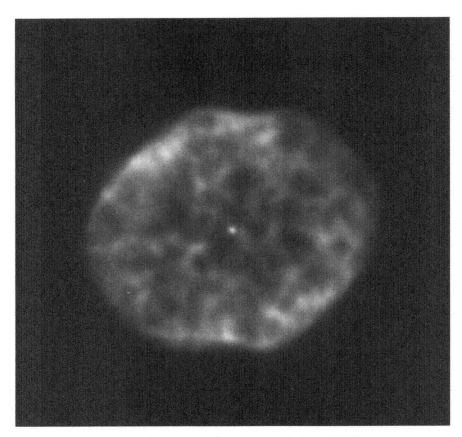

Figure 7.8. The central star of the planetary nebula NGC 1501 is a PG 1159-type variable with a 0.05 magnitude amplitude and a 20-minute variation period. Its spectrum also shows it to be a high-temperature "O VI star," one that radiates powerful emissions of five-times-ionized oxygen. [AURA/NOAO/NSF, Y. H. Chu and G. H. Jacoby.]

missed, coming in at magnitude 1.8. Dimmer (more distant or more obscured by interstellar dust) "new stars" are much more common, several faint novae found every year. Allowing for the obscuring effects of thick clouds of interstellar dust, about 30 take place annually within our Galaxy.

Astronomers have rarely seen the stars before they erupted, but they can easily locate them afterward following their return to normal. What they find is a huge jump of 10 magnitudes or more – a factor of 10,000 in visual brightness – that occurs over only a day or two. Novae may for a time become as luminous as supergiants, with absolute magnitudes approaching −10: they are, as might be expected, erupting at the Eddington limit, where the luminous blue variables reside. But their glory is short-lived. After a month or two they typically drop by 3 magnitudes or so, and after a few years (or sometimes decades) return to their original states. Then a few years after the eruption we see clouds of gas expanding around the stars, visible evi-

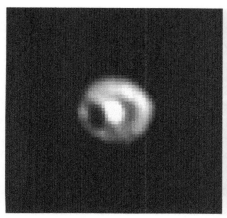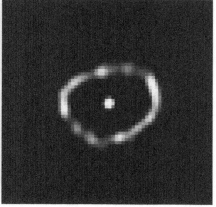

Figure 7.9. Two pictures taken 8 months apart (in 1993 and 1994) with the Hubble Space Telescope show the shell surrounding Nova Cygni 1992 expanding from around 800 to 1000 AU. [F. Paresce and R. Jedrzejewski, STScI, and NASA/ESA.]

dence of mighty blasts that have hurled matter out into space at speeds of thousands of kilometers per second.

The interpretation of the nova phenomenon is one of the greater success stories of twentieth-century astronomy. A nova is the product of two stars in a close double system, one an ordinary low-mass main sequence star, the other a white dwarf, the duplicity revealed by eclipses. If the stars are sufficiently close, they will also raise mutual tides. The Earth and Moon provide a good example of tidal behavior. The side of the Earth facing the Moon is closer than that facing away, so the gravitational force is stronger on the facing side. As a result, the Moon exerts a stretching force on the Earth that causes the Earth to be slightly elongated along the direction to the Moon (by a few centimeters) and, more familiarly, makes the Earth's waters flow toward the same line, creating a pair of watery bulges under which the Earth rotates. The Earth creates an even larger tide in the Moon.

As gaseous bodies, stars can be considerably stretched by a close companion, and can even have mass torn away. Large orbiting bodies in close proximity are surrounded by a three-dimensional surface in which the acceleration of gravity is effectively zero, the result of the stars pulling in opposite directions and of the centrifugal force caused by orbital revolution. In a double-star system destined to become a nova, the larger and more tenuous main sequence star is stretched by a tide until it fills this "zero-gravity" surface and thereby spills matter toward the white dwarf. The mass does not flow directly to the degenerate star, however, but first into a surrounding disk. Hydrogen-rich mass is then fed from the disk onto the surface of the small star, various instabilities in this "accretion disk" causing the unresolved pair to flicker.

The huge surface gravity of the white dwarf compresses and heats the fresh hydrogen falling upon it. The hydrogen eventually – and suddenly – ignites as an

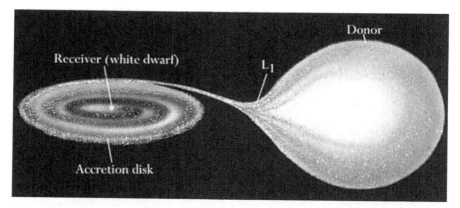

Figure 7.10. In a close double star, tides can cause the larger of the components to fill an enclosed volume at whose surface the gravity of the stars is equalized, allowing mass to flow outward through the point labeled "L1." The matter passes first into an accretion disk and then on to the other star's surface, here a white dwarf. [From *Stars* by J. B. Kaler © 1992 by Scientific American Library. Used with permission of W. H. Freeman and Company.]

uncontrolled nuclear bomb operating by the carbon cycle. The explosion, which can be seen clear across the Galaxy, expands the surface of the white dwarf to the Eddington limit, causing it to expel mass. In one view, the orbiting main sequence star helps stir the ejected matter and drive it outward at high velocity. The high temperature remaining on the white dwarf subsequently illuminates the expanding nova remnant, allowing us to see the debris of the detonation some time thereafter.

If the two stars of a binary system are so close as to generate a nova, they should have disrupted each other when both were on the main sequence. They must therefore have started out farther apart, and over time have been drawn together. Begin with two reasonably separated stars, one more massive than the other. The bigger one starts evolving first, ultimately growing into an asymptotic branch giant. If the two were initially sufficiently close, the outer layers of the giant star will become tidally disrupted by its lower-mass companion; the giant then throws matter toward the dwarf. So much can be lost so fast that the mass spills outward, and the giant can embrace its companion in a common envelope. Friction slows the orbital motion and the stars spiral together. Such a picture can help explain the severe bipolar structures of some planetary nebulae, the bodies produced by the evolving giants. By the time the giant has created its planetary nebula and evolved into a white dwarf, the stars can be quite tightly linked, and the stage is set for the creation of a nova.

After the surface of the white dwarf is blasted away by the nova, the phenomenon can repeat itself. A properly configured double star might go off every hundred thousand years or so. If conditions are just right, some – recurrent novae, like RS Ophiuchi – explode every few decades or so. In between times, instabilities in the accretion disks are believed to produce minor events that mimic novae, the so-called "dwarf novae" like SS Cygni and U Geminorum, which pop off by a few magnitudes every month.

If the stars are not quite so close, interaction can occur farther down the line when the originally less-massive star begins its evolution. This time a giant, possessed of a strong wind, can pass matter to the white dwarf via an accretion disk. The point where the flow hits the disk can be heated to very high temperatures and can radiate so much at high energies that the flow becomes ionized. Now the observer sees an M (or other) giant whose spectrum contains emission lines that look like those of a planetary nebula, but with no obvious nebulosity. The combination of two very different characteristics, one from a hot star, the other from a cool one, led astronomers to call these creatures "symbiotic stars." Over 100 are known. Adding to the complexity, quite a number contain Mira variables, such systems a welding of the large and the small, of the cool and hot.

On occasion a symbiotic star will undergo a nova-like eruption that can take years to settle down. At its peak, CH Cygni, at fifth magnitude, was visible to the naked eye. R Aquarii, a Mira with a white-dwarf companion, erupted in the 1920s, the outbursts surrounding the star in a complex expanding nebula. Some outbursts are probably caused by thermonuclear runaways fueled by the mass accreted by the white dwarf; in other cases they are caused by great instabilities in the accretion disks. Little about them is clear. A few symbiotics even have planetary nebulae around them with extreme bipolar shapes that seem to result from the stirring of the giant by the white dwarf, providing credence for theories that suggest that all bipolar planetaries are the creations of binaries.

## Magnetic fields

One to two percent of white dwarfs enlarge upon their already curious characteristics with astonishing magnetic fields. Stellar magnetism is relatively common in stars. The Sun, for example, has a weak polar-aligned field that is about twice the strength of Earth's and that helps drive the 11-year solar activity cycle. Far down the main sequence, dwarf M flare stars tell of even greater magnetic activity. Magnetic fields also play a prominent role upward along the main sequence, where among the A stars we find a special set with peculiar spectra and odd chemical compositions (called Ap stars) that exhibit magnetic fields with strengths up to 30,000 times that of the Sun.

Magnetic fields are detected through the Zeeman effect, which causes spectrum lines created within the

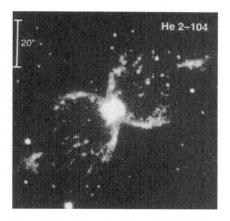

Figure 7.11. (See also Plate XVIII.) The "Southern Crab" is a planetary nebula with a symbiotic star at its center, and is the result of mass loss stirred by a hot white dwarf. [AURA/NOAO/NSF, J. H. Lutz.]

Figure 7.12. A Type Ia supernova explodes in M63, signaling the end of a white dwarf in a double-star system. [University of Illinois Prairie Observatory.]

field to be split into multiple parts. In main sequence stars the splitting is subtle, at most under an ångstrom. In contrast, the Zeeman splitting produced by the magnetic white dwarfs can spread the components of a hydrogen line through most of the optical spectrum, revealing magnetic fields up to 300 million times more powerful than the Sun's. Apparently, ordinary stellar magnetic fields are squeezed down and intensified in parallel with the stars' increased densities and gravities. But we have no idea of why only a few white dwarfs possess them and the rest do not.

The magnetic fields of white dwarfs within interacting binaries can be so strong that they affect the mass lost from the companions, causing the gas to flow onto the magnetic poles, resulting in peculiar flickering behavior. Nova Cygni 1975 involved a magnetic white dwarf, the magnetism probably influencing the shape of the ejected nebula.

## Supernovae

The biggest and smallest stars have a remarkable common property: both produce supernovae. Supergiants that evolve iron cores are the progenitors of Type II supernovae, and there seems little doubt that the Type Ib variety are made by supergiant hydrogen-less Wolf–Rayet stars. But Type Ia supernovae, those without hydrogen in their spectra, the ones that can occur in galactic halos where there are no massive stars, have yet to be addressed. Their halo residence indicates that they must somehow involve lower-mass stars. To understand such blasts, we must find a means of concentrating the stellar energy. The most obvious way is the evolution to the dense white-dwarf state.

The situation is not as clear as it is for Type II supernovae, and several possibilities are argued, all involving white dwarfs in double systems. In a popular picture, a massive white dwarf lies near the Chandrasekhar limit. Matter fed from the main sequence star can still produce a nova, but if sufficient mass can be accreted before the nova ejects it back into space, the white dwarf can be pushed over the limit. Degeneracy can no longer hold gravity back and the white dwarf suddenly collapses, the heat causing a sudden onset of searing nuclear fusion and a devastating explosion that quite possibly destroys the white dwarf altogether. Such detonations are even brighter than the massive-star variety, going up to absolute visual magnitude −19.

Support for this scenario comes from the similarity of Type Ia supernovae that are seen in other galaxies, suggesting that the initial conditions cannot differ by much. In turn, their similarity makes Type Ia supernovae excellent distance indicators for very distant galaxies. When we see one go off, we know its peak absolute magnitude. Measurement of its apparent magnitude gives the distance and such interesting numbers as the expansion rate, age, and fate of the Universe.

A second possibility again begins with the standard picture in which a white dwarf and ordinary dwarf orbit each other. The ordinary dwarf evolves to gianthood and encloses the white dwarf in a common envelope that causes the two to spiral even closer. The end product is an intimate pair of white dwarfs locked tightly in gravitational embrace. Gravity is best explained through Einstein's theory of relativity as a bending of spacetime (the four dimensional construct of space and time in which we appear to live) by mass. We sense the distortion by falling downward toward that mass. Acceleration of the mass should produce waves that carry away energy. Since orbiting bodies are continuously accelerated, they must lose energy through gravitational waves, causing the bodies to spiral closer together. The effect on the Earth and in most double stars is insignificant. But two close white dwarfs can so distort spacetime that their gravitational radiation over a long period of time might make them merge and exceed the Chandrasekhar limit. The result again is nuclear devastation. A third theoretical possibility involves runaway carbon ignition in accreting white dwarfs that are below the Chandrasekhar limit.

Though the exact process that makes Type Ia supernovae is yet to be understood – and in fact more than one may actually operate in nature – there seems to be little doubt that white dwarfs are involved.

## The end (at least of white dwarfs)

Students of astronomy are barraged with the fact that stars are gaseous. Surprisingly, there are notable exceptions. White dwarfs seem destined to cool forever, moving ever lower on the HR diagram along the white-dwarf sequence. A white dwarf's matter is in such a dense state, however, that as its surface temperature drops toward the bottom end of the white-dwarf sequence, below about 7000 kelvin, it begins to crystallize and turn solid.

But though these little stars continuously cool, they lose most of their heat by slow conduction, and the rate of cooling is terribly slow. In the entire history of the Galaxy, no white dwarf – not even one created in the Galaxy's earliest days – has yet had time to move off the HR diagram and become invisible. There are no "black dwarfs." Therefore there must be a visible end to the string of such stars. If we can find it we can examine the end-products of the earliest generations of stars, and if we know the evolutionary and cooling rates from theory we can measure the age of the Galaxy.

These dim reddish stars, just under 4000 K and absolute visual magnitude 17, are hard to locate and study. From the numbers of the faintest white dwarfs found in the populous Galactic disk, astronomers estimate the disk's age at about 10 billion years, which is consistent with other determinations. In principle, the technique can be extended into the Galaxy's older halo, which from the ages of globular clusters seems to be around 15 billion years old. But stars are sparse in the halo, distances great, and the few stars to be found there visually very dim. Though the statistics on halo white dwarfs have yet to be compiled, they will someday provide a powerful means of checking the age of the oldest part of the Galaxy and for determining something of its ancient development. The white dwarfs, at the end near-dead frozen stars, have significance all out of proportion to their tiny sizes and faint glows.

## Neutron stars

As small as white dwarfs are, the squeezing room within an atom is so great that they are not even close to being the smallest stars possible. At a mass of 1.4 times that of the Sun, electron degeneracy no longer provides support, and the white dwarf can no longer maintain itself. But if electrons can become degenerate, filling their little six-sided boxes, so can more massive particles, like neutrons. But for these to become degenerate, they must be squeezed much tighter, almost to nuclear density itself. Protons could become degenerate too, but protons are always found in company with electrons and, at the densities required, the protons merge with the electrons and are forced to become neutrons. The degenerate neutron star therefore becomes the only choice, and is the only stellar successor to the white dwarf.

Since only about one part in $10^{15}$ of the atom is actually occupied by matter, and since the average density of the Sun is about that of water (one gram per cubic centimeter), we could contract a star to a density of about $10^{15}$ grams per cubic centimeter; neutron degeneracy is actually reached at an average of $10^{14}$ (a hundred trillion) grams per cubic centimeter. At that density, the Sun would have a diameter of only 30 kilometers. Actual neutron stars are thought to be only somewhat smaller, closer to 20 kilometers. Conditions are so extreme that neutron-star surfaces crystallize.

How do you collapse a white dwarf and not tear it apart, as seems to happen in a Type Ia supernova? Massive stars have the answer: collapsing iron cores near the Chandrasekhar limit that produce Type II supernovae, cores that cannot make white dwarfs. In their collapse, in which the iron is broken back to its fundamental particles, which in turn make neutrons, the cores quickly reach the degeneracy level. The pressure of these degenerate neutrons then stops the contraction, and a neutron star in born.

Neutron stars were conceived as the remains of supernovae in the 1930s by Fritz Zwicky of the Mt Wilson Observatory. Even an odd star at the center of the Crab Nebula, the gaseous remnant of the supernova of 1054, had been tentatively identified as one, as it had no absorption lines. But it took the science of radio astronomy and the discovery of pulsars in 1967 to reveal them and their bizarre natures.

The story is a famous one. The British radio astronomer Anthony Hewish had set up a specialized radio telescope to monitor the flickering of distant radio sources as their radiation passed through the corona of the Sun, the purpose to examine the corona's properties. Instead, his then-graduate student Jocelyn Bell began to record faint but unmistakable "pulses" of radiation spaced 1.3 seconds apart. Though the pulses might disappear for a time, they maintained a perfectly regular period of 1.337011 . . . seconds. Hewish and Bell first considered the wild possibility of having discovered an interstellar communications beacon, but the discovery of more of these strange sources convinced astronomers that they must be natural.

Only two mechanisms could produce such pulses, actual pulsation and rotation. Nothing seemed to be able to pulsate that fast, that regularly, certainly not a white dwarf. A white dwarf spinning at such a speed would also tear itself apart, so the body had to be very small. The pulsing source, this "pulsar," must be a rotating neutron star. Shortly thereafter, a radio pulsar with a then-astonishing period of 0.033 seconds was found in the Crab Nebula and later identified with Zwicky's suspected neutron star. Neutron stars are real, the young one in the Crab, a star of well over a solar mass only a few tens of kilometers across, rotating 30 times per second.

Though all the details of pulsar theory have yet to be worked out, the basic mechanism is surprisingly simple. Pulsars have powerful magnetic fields that have been squeezed down with the star until they too have reached enormous densities, their strengths some $10^{12}$ times that of the Earth or Sun and 10,000 times greater than the fields of the most extreme white dwarfs. Like most astronomical magnetic fields, a pulsar's magnetic axis is tilted relative to its rotation axis. The spinning magnet creates electric currents that radiate their energy outward along that wobbling magnetic axis. The pulsar therefore radiates something like a lighthouse, and if the Earth happens to be in the way of the spinning beam, we will see the neutron star briefly light up, the effect a "pulse" of radiation.

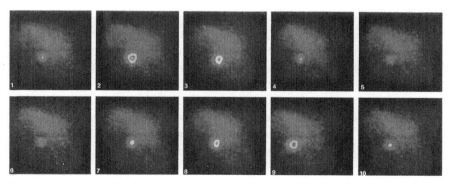

Figure 7.13. (See also Plate XIX.) The pulsar in the Crab Nebula, here radiating X-rays, turns on and off 30 times per second. The flash lasts for only a few-thousandths of a second. [F. R. Harnden, Center for Astrophysics.]

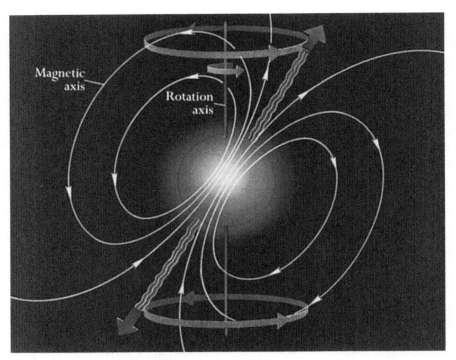

Figure 7.14. A pulsar spins wildly in space, beaming out radio (or even optical, X-ray , or gamma-ray!) radiation along a tilted magnetic field. If the magnetic axis points to the Earth during the gyration we will receive a burst of radio energy. [Based on a drawing by G. Greenstein in *Frozen Star*, Freundlich, New York, 1983, art from *Stars* by J. B. Kaler © 1992 by Scientific American Library. Used with permission of W. H. Freeman and Company.]

Hundreds of pulsars, with a great variety of periods, are known. The light curves are fascinatingly complex. Some pulsars have quite notable interpulses halfway between the main pulses that are produced by radiation from the other pole. The pulses also have substructures that change continuously; yet the characteristic shape found by averaging the substructures over long periods of time is reliably constant. For unknown reasons, the pulses occasionally disappear, sometimes for days or weeks. But when they return, they come back right on schedule.

The arrival times of the pulses change with radio frequency (or wavelength), a consequence of their passage through the ionized gases of the interstellar medium. This thin gas delays the signals, their arrival times becoming later as the wavelength shortens. A comparison of the arrival times at two frequencies coupled to an estimate of the average gas density in interstellar space then leads to the distance. Conversely, the delays can be used to examine the state of the interstellar gases.

A young pulsar is born with high rotation speed, allowing it to radiate at high energies as well as low: the Crab pulsar, for example, pulses not only optically but even in the X-ray and gamma-ray parts of the spectrum. But the radiation must take its toll. A pulsar's energy is derived from its magnetic field and ultimately from its spin, exactly what we observe, though sometimes one can be contrary. Like cool white dwarfs, pulsars have interior crystalline structures that during the spin–down occasionally adjust themselves, causing a contraction of the star and a slight increase in rotation speed. As the neutron star slows, it loses its ability to generate high-frequency radiation and becomes visible only in the radio. By the time the period is up to 4 seconds or so, it is too old and weak to radiate much at all, and it disappears from view. Old non-pulsing neutron stars must be all around us, radiating like hot black-bodies, but their surface areas are so small that they are very difficult to see. There must be hundreds of millions of them accumulated from a Galaxy-lifetime of massive-star supernovae.

Such a non-pulsing neutron star, found first by its X-ray radiation, was finally imaged by the Hubble Space Telescope in 1998. The star's apparent visual magnitude is around 25 (though perhaps fainter), and its position in front of an interstellar dust cloud suggests a distance less than 400 light-years. (If the star were in back of the cloud, it would not be visible.) Adopting these rough values leads to an absolute visual magnitude around 20. The temperature is estimated from the observations to be a bit over a million degrees kelvin, the high value suggesting that it is probably not so much an old pulsar but one that is just turned in the wrong direction for us to see. Though seemingly faint, the high temperature leads to an immense bolometric correction of some 13 magnitudes (considerably greater than for planetary nebulae). If we could see all the radiation with our eyes, this neutron star would shine at absolute (bolometric) magnitude +7 and be visible with small binoculars.

Oddly, though pulsars are clearly produced by supernovae, the relation between pulsars and supernova remnants is not great. Most pulsars are devoid of surrounding gas and most supernova remnants have no pulsars. The supernova remnant of course could dissipate, leaving a "naked" pulsar. A pulsar buried within a remnant could also just be oriented the wrong way. Another explanation is offered by the high proper motions that pulsars can exhibit. From their motions and distances, we find that they can speed along at hundreds of kilometers per second, far faster than the massive stars that fostered them. The most ready explanation is that a core–collapse supernova event and resulting explosion can be slightly off-center, which gives the resulting pulsar a huge kick, enough to

Figure 7.15. (See also Plate XX.) The Hubble Space Telescope reveals a tiny non-pulsing neutron star (*arrow*) radiating by virtue of its 1.2 million degrees kelvin temperature. [F. Walter (SUNY at Stonybrook), STScI, and NASA.]

Figure 7.16. The Vela X-1 pulsar pushes a shock wave in front of it as it speeds through the interstellar medium. Moving at 90 kilometers per second, it ejected itself from the Vela OB1 association some 2.5 million years ago when the supernova that produced it exploded off-center. [Danish Telescope, European Southern Observatory, L. Kaper *et al.*]

shoot it out of its decelerating remnant and even to loft it right out of the disk of the Galaxy.

Most pulsars, for unknown reasons, seem to be single stars, without companions. A small number, however, are indeed in binary systems, recognized by small variations in the pulse arrival times. As an orbiting pulsar comes at us, its pulses increase in frequency, as each burst of radiation occurs when the pulsar is a bit closer and the light travel-time is less. Conversely, when the pulsar is receding in orbit, the pulses are delayed, that is, we see something akin to a Doppler effect. From the pulse variations, astronomers can construct orbital properties of the stars. The companion to the first such double discovered is either another neutron star or a massive white dwarf near the Chandrasekhar limit. These two are locked in such extreme embrace that they must radiate relatively powerful gravitational waves. Since gravitational radiation removes orbital energy from the system, the orbital period should slow. And so it does, right on schedule with the rate predicted by Einstein's relativity, both supporting the famous theory and providing support for the merger theory of white-dwarf supernovae.

## Rebirth and destruction

Stars are so tightly gravitationally bound that a supernova will not disrupt a companion, and the duplicity survives. If a white dwarf can interact with a main sequence star, so can a neutron star, but with very different, and bizarre, results. If the two are close enough, mass is again transferred from the ordinary dwarf to the neutron star via an accretion disk. But because of the intense gravity of the neutron star, the disk is compressed to such high density and temperature that X-rays are produced. Several such X-ray binaries are observed by Earth-orbiting satellites.

The neutron star within an X-ray binary will accumulate the hydrogen that is falling vigorously from the accretion disk. The fresh hydrogen will fuse to helium by means of the carbon cycle, and when enough helium has accumulated at the base of the new layer it violently fuses to carbon, producing an intense burst of X-rays that can last for a minute or two. Typical X-ray "bursters" are quite active, repeating the cycle every few days or even hours.

Three stranger things are yet to come. Matter falls from the accretion disk onto the neutron star at a glancing angle with such force that the neutron star begins to spin faster, much like a person continuously striking a bicycle wheel with one sideways blow after another. As the rotation speed picks up, the old neutron star begins once again to act like a pulsar, but only after it is spinning hundreds of times per second, the pulse period now in thousandths of a second, in milliseconds. To the eye, such a star would be simply a blur, the pulses coming so fast that if you could hear them they would sound like a musical tone, a celestial violin string singing near middle C.

The Black Widow pulsar is by chance an eclipsing double, from which we can tell the nature of the system. The pulsar's companion has a radius 0.2 that of the Sun and an extraordinary mass of only 0.02 solar, far below anything an ordinary dwarf

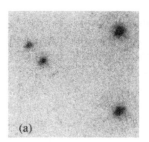 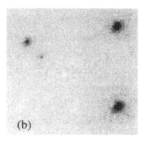 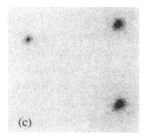

Figure 7.17. In this negative image, the "Black Widow" millisecond pulsar (observable only in the radio) is periodically eclipsed by an optically visible companion (at upper left) that is being eroded away by the pulsar's radiation. (a) The pulsar is in front of the companion and we see the companion's heated surface. (b) The heated surface of the companion turns away from us and the star dims. (c) The heated surface of the companion is now turned away from us. Its other side is so dim that it too disappears. [A. S. Fruchter *et al.*, STScI, and NASA.]

should have. The only explanation that comes to mind is that the immense energy radiated by the millisecond pulsar is eroding away its companion, explaining why some millisecond pulsars, in contradiction to theory, have no companions: they have destroyed them.

The debris, however, seems to be able to take on a new existence. Regular variations in pulse arrival times reveal that a pair of millisecond pulsars are being moved slightly back and forth by orbiting bodies that have masses and orbital radii comparable to those of our terrestrial planets. Apparently, some of the mass of the eroded star re-accumulates as smaller solid bodies. No life would be possible because of the hugely energetic natures of the systems, but the pulsar planets at least demonstrate that planets can form out of almost anything, an important consideration for the next chapter.

## Ultimate contraction . . .

White dwarf – neutron star – then what? Electron degeneracy has a limit, and so does neutron degeneracy. If the mass of the dense star exceeds two or three times that of the Sun, even neutrons packed to the density of nuclear matter cannot hold back the inexorable force of gravity, and the star must collapse again. But this time there is no salvation. There is nothing left to stop the squeeze, and the star collapses forever. As it contracts it passes a radius at which the escape velocity (the speed at which a body must fly away never to return) hits the speed of light. At that point light, radio, and all the other forms of energy, can no longer get away, and the object (can we still call it a star?) disappears from view.

The result is the storied black hole. Its gravity is still there, but our attempts to view it are futile, as it is in a sense – except for its gravity – gone from space. Such a body can really be described only by the theory of relativity, in which we consider gravity as a distortion of spacetime caused by mass. A black hole's gravity is so great that it literally creates a puncture in spacetime, an infinitely deep well from which nothing, not even light, can depart.

Though black holes themselves do not radiate light, their surroundings can. We can detect one by its effect on a stellar companion, either by Doppler shifts induced in an orbiting neighbor or by the accretion of matter. If the companion swells in its evolution or approaches too close to the black hole, tides can distort the normal star and cause matter to flow from it to an accretion disk, whereupon it passes into the hole in spacetime never to be seen again. The accretion disk, however, is so hot that it radiates X-rays. But so do accreting neutron stars. We can presumably tell the difference by measuring the mass of the collapsed body by the orbital effect it has on its more-normal companion. If the mass is above the neutron-star degeneracy limit, we can at least presume we have netted a black hole. There are several candidates in our Galaxy and in the Large Magellanic cloud. Among the best is Cygnus X-1, in which an invisible body with a mass estimated to be well above the neutron-star limit, orbits a typical B0 supergiant. B supergiants do not radiate such X-rays, and

Figure 7.18. Cygnus X-1 (*arrow*), set here in context in its constellation (Deneb is at upper left, Albireo at lower right), is a powerful X-ray source. The X-rays are most likely caused by matter flowing from a B supergiant into an accretion disk surrounding a black hole. [Author's photograph]

its combination with an invisible companion strongly suggests the presence of a black hole. Other candidates have masses that hover around six times that of the Sun.

We are fairly sure that black holes can develop from supernovae, but we are not yet certain why some produce them and why others make neutron stars. From what little evidence we have, it appears that black holes result from higher-mass progenitors, and that the more populous lower-mass O stars (up to about 20 solar masses) make the neutron stars and pulsars. From the number of X-ray black-hole binaries we see, the Galaxy may contain millions of invisible black holes. Though Bessell started us on the road to understanding the smallest stars over 150 years ago, fascinating mysteries remain that will keep scientists working for years to come. But in a sense, these are the details. We at least now know that one extreme leads to another, the biggest stars becoming the smallest. Moreover, the bigger they are, the harder they fall, giants turning into white dwarfs, supergiants into neutron stars, bigger

169

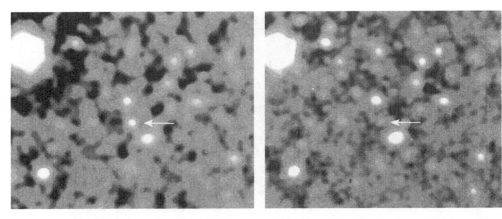

Figure 7.19. A flash of light (*left*, *arrow*) associated with a gamma-ray burst explodes from a distant galaxy, seen as a faint smudge (*arrow*) on the right. The pictures are only half-a-minute of arc across. [S. G. Djorgovsky and S. R. Kulkarni, Caltech/W. M. Keck Observatory.]

ones into black holes, all the result of the inexorable force of gravity, which more than anything else drives the lives of the stars.

## . . . and ultimate violence

At least once per day an Earth-orbiting gamma-ray telescope observed a bright gamma-ray burst. Such bursts can come from any part of the sky irrespective of Galactic position. Explanations ranged all over the theoretical map until the bursts were identified with distant galaxies, which ruled out local causes and revealed the bursts as the most violent events known in nature, up to 100 times more luminous than an "ordinary" supernova. There is still no accepted theory. For some time the bursts were widely believed to be caused by neutron-star mergers, representing vastly amplified versions of the white-dwarf merger theory of Type Ia supernovae. But even these fall short of the greatest energies observed. Also popular is a "hypernova" theory in which a spinning stellar core collapses into a black hole, dragging yet more matter with it. Yet another idea is that some millisecond pulsars may rotate much faster inside than they do at their surfaces, resulting in a wrapping and concentration of the magnetic field to a hundred thousand times that found in a normal pulsar. When the magnetic stored energy rises to a critical level, it is released, creating the immense burst.

Similar violence is found here. On August 27, 1999, Earth was pounded by a blast of X- and gamma rays that saturated the detectors of spacecraft and partially ionized the upper atmosphere. Their origin was an unusual pulsar – now called a "magnetar" – that has a magnetic field $10^{14}$, a hundred trillion, times that of Earth and 100 times that of a normal pulsar. The powerful field is thought to have cracked the pulsar's crust to produce the immense burst. The magnetar is 25,000 light-years away. Even day-to-day life is not isolated from the Galaxy, but very much a part of it.

# Chapter 8

# The youngest stars

The stars seem ageless. If you could go deep into time, thousands of years ago, to ask your ancient ancestors about the sky and the constellations, they would describe almost exactly what you see today. Upon first hearing it therefore comes as a surprise to learn that stars – like people – are born, age, and die. We have looked closely at the extreme products of stellar aging: at the coolest, the brightest, the hottest, the biggest stars; and at the end-products of that evolution: at the smallest stars, the white dwarfs, neutron stars, and black holes. But little has been said about origins. Turn now in a different direction, from physical condition to time, to age, to the beginnings of stars, to extreme stellar youth.

## Start with the Sun

How can we know stellar ages when, except for catastrophe, we hardly ever see the stars change? We cannot actually watch a star be born, go through its youth and middle age, and finally die. We have to catch individuals along the way and string them together with the aid of logic and theory and the fundamental reference, the Sun and its family. From the nature of the Solar System – that the planets all orbit in the same direction as solar rotation, all approximately in the

Figure 8.1. (See also Plate XXI.) Earth and Sun are partners, born at nearly the same time, 4.5 billion years ago. [Author's photograph.]

solar equatorial plane – we have long surmised that the planets and the Sun were born at the same time. We then need only know the age of the Earth or the other small bodies of the Solar System, to which we have direct access, to find the age of the Sun.

There is a long history to determining the terrestrial age, starting with the study of Biblical literature that suggested that our home is only 6000 years old. Then scientists began finding clues that led to much greater ages. From the rates of formation of geological deposits, and for the sea to have reached its present salinity, the Earth ought to be at least 100 million years old and perhaps much older. The final answer came from the discovery of radioactivity by Henri Becquerel in 1896. He had accidentally placed a photographic plate next to a sample of pitchblende, an ore of uranium. When he developed the plate, he found it fogged – the mineral was emitting radiation.

We are used to the chemical elements being permanent. But many are not. Isotopes of elements that contain too many, too few, or just the wrong number of neutrons can be unstable and will decay into something lighter with the release of radiation, rendering them "radioactive." Hydrogen (one proton) with one neutron (deuterium) lasts forever, but add another neutron to make $^3$H – tritium – and in a few decades it will be gone. Technetium, element number 43, has no stable isotopes at all. Its presence in evolved asymptotic branch giant stars clearly demonstrates that chemical elements can be made within stellar furnaces.

The heaviest elements, those beyond bismuth (element 83), have no stable isotopes. Such substances as radium, thorium, and uranium are inherently unstable and constantly radiate energy in the form of particles and real electromagnetic radiation. If you get too close to them for too long, they are lethal. Under natural conditions, a radioactive element does not decay instantly, but by a strict timetable given by the element's "half-life," the time it takes for any given amount of the substance to cut itself in half. Half-lives differ enormously. A kilogram of $^{238}$U will decay into half-a-kilogram in 4.5 billion years, and the longest-lived version of thorium ($^{232}$Th) takes over 10 billion years to do the same thing. In spite of their steady decay, this time period is so long that these isotopes are effectively permanent parts of the Earth's crust. But $^{235}$U, used in atomic bombs, takes less than a billion years to decay in half, and the most stable isotope of radium, $^{226}$Ra, takes a mere 1600 years, rendering it very dangerous.

The phenomenon provides natural clocks. $^{238}$U decays through a series of elements (including radium) into lead, $^{206}$Pb, while $^{232}$Th eventually turns into $^{208}$Pb. A variety of other isotopes are available as well. When a rock solidifies from the liquid state, from magma inside the Earth (ejected through a crustal crack or volcano), it seals in then-current ratios of the abundances of the daughter products of radioactive decay to those of the parents. The amount of the daughter thence steadily increases at the expense of the parent. A measure of the ratios of the abundances of the daughters to parents subsequently gives the rock's age.

A good example of the process, one used in archaeology, involves radioactive $^{14}$C. The isotope is produced naturally in the Earth's atmosphere by cosmic rays –

the high-speed particles from space produced in supernova blasts – and is absorbed by living things. When a plant dies, it seals in a known ratio of $^{14}C$ to $^{12}C$. A measure of the ratio then gives the time since the plant's death and an excellent clock for dating old wood ashes from ancient campfires and the ages of ancient pottery shards.

Such radioactive dating methods applied to the Earth's rocks give a maximum age of about 3.8 billion years. That, however, is the age since solidification. The Earth could have been molten for a long time before that, or the early rocks somehow destroyed. If that is the case, to reach back further we have to look at other, smaller bodies in the Solar System that should have solidified much faster, hence earlier. The Moon is smaller and colder than the Earth. The oldest rocks brought back by the Apollo astronauts are 4.51 billion years old, and the most ancient meteorites are dated at a maximum of 4.53 billion years. Since we can find nothing older, we identify this figure with the age of the Solar System and with that of the Sun, a figure confirmed by the application of the theory of stellar structure to the Sun itself. The fossil record shows that the Sun has not changed much over that period of time. Scientists therefore had to find a mechanism that could have kept that great body shining for so long.

In 1926, Sir Arthur Eddington had the beginnings of the answer. An atom of helium, which contains four nuclear particles, was known to weigh 0.7 percent less than four hydrogen atoms. Reasoning from Cecilia Payne's discovery that stars are nearly all hydrogen, Eddington surmised that somehow four protons fuse into helium and release their mass as energy via Einstein's famous relation $E = mc^2$. The speed of light, $c$, is so great that only a tiny amount of mass is needed to produce a staggering amount of energy: a thousand-megawatt power plant could in principle be run for a week on a kilogram of hydrogen. In 1940, Hans Bethe and Charles Chritchfield figured out the details of the mechanism, recognizing the proton–proton chain, which can proceed only at the high temperature of the solar core (15 million degrees kelvin; the minimum required is about 7 million). From the known rate of fusion and the size of the fuel supply, astronomers could then calculate that the Sun should have a 10 billion-year hydrogen-fusing lifetime. The Sun is therefore about half-way through its main sequence lifetime.

Though the Sun is not young, it provides us with physical and temporal benchmarks from which we can explore the chronology of the Universe and

Figure 8.2. This seemingly drab meteorite (a "carbonaceous chondrite" from the Allende fall in Mexico), about two centimeters across, is an example of the oldest things known. Such rocks, actually small primitive asteroids that have hit the Earth, are used to date the Solar System. [Author's photograph.]

from which we can search for true stellar youth. Since the Sun is an ordinary main sequence star, it is reasonable to assume that the main sequence is a band of stars that all run on hydrogen fusion. Studies of double-star orbits, which allow measurement of stellar masses, show that the brighter stars are the more massive. The major difference among main sequence stars is that above about two solar masses, hydrogen fusion proceeds mostly by the carbon cycle (also discovered by Bethe), wherein carbon atoms aid in the fusion process. Theory demonstrates that stars in other parts of the HR diagram are the products of stellar evolution, the giants and supergiants created when the main sequence hydrogen fuel finally ran out. These evolved stars use other, shorter-lived fusion reactions to help create energy. A huge number of yet other reactions make the known chemical elements during ordinary evolution and during the explosions of supernovae. The great blasts, and the winds from lesser stars, seeded the interstellar gases with the heavy stuff from which the Earth was eventually made.

## Birthplaces

The greater the mass of the star, the faster the fuel is used and, as a result, high–mass stars live shorter periods of time than do those of lower masses. Since high-mass O stars do not live very long, and cannot move far from their birthplaces, we can recognize truly young groups of stars, clusters and associations, by the O stars' presence. Even a casual glance at the broken structure and apparent holes and gaps in the Milky Way shows that dark clouds swarm through space, their dust blocking the light of more distant stars. Spatially allied with the dark clouds are the hosts of diffuse nebulae that contain O stars, in turn relating the O stars to the dusty clouds of interstellar space. Moreover, the cooler, though still young, B stars are commonly related to reflection nebulae, whose dust scatters the starlight. Dust clouds must therefore be the birthplaces of stars, even if no O or B stars are present. We look there and find new stars being born as we watch.

Stars seen through the dust are both dimmed and reddened, as the blue component of the starlight is extracted (absorbed and scattered) by the dust more readily than the red component. From the way in which the starlight is reddened, we learn that the particles are tiny, typically much less than a thousandth of a millimeter across. On the average there is far less than one grain per cubic meter of space set within a gas whose density averages about one atom per cubic centimeter. But since the grains clump into denser clouds (with gas densities upward from 1000 atoms per cubic centimeter), and since the path lengths through with we look are so great, this small amount has a powerful cumulative effect that can entirely prevent us from seeing stars behind the denser clouds. About one percent of the mass of interstellar space is in the form of these little particles.

The clouds also superimpose their absorption spectra on the stars whose light can penetrate to the Earth. We find hosts of narrow absorption lines from common metals, and in the infrared broader absorptions from solids made of silicates. In the

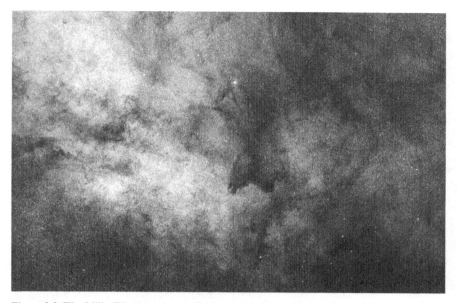

Figure 8.3. The Milky Way (seen here in Sagittarius) is filled with dark dusty clouds that hide the background stars. [Palomar Observatory, California Institute of Technology.]

ultraviolet there is also a broad absorption believed to be produced by amorphous carbon (perhaps something like graphite), and in the infrared the signature of crystallized carbon, of diamond dust! We know from stellar studies that ordinary oxygen–rich Mira variables eject vast quantities of silicate dust into the cosmos, and that carbon stars expel massive amounts of solid carbon, so much so that the stars can be temporarily buried within the sooty winds. These evolving giant stars produce most of the solid grains needed to make the dust of interstellar space. Once released into the cold, the dust grains become more complex, absorbing metals and coating themselves with ices, as revealed by spectra and by the fact that the interstellar gas is notably depleted in a variety of heavy elements, those that have condensed onto the dust particles. The diamonds are probably created by the pressures produced by the shock waves from exploding stars, from supernovae.

Over the past 30 years, radio astronomers have found that the dust clouds are actually the visual manifestations of vast clouds of molecules. The largest, the "giant molecular clouds," are among the most massive units to be found in the Galaxy. They are made mostly of molecular hydrogen, $H_2$, but are loaded with a vast variety of other molecules. Some 100 species are known, from simple things like carbon monoxide, water, and ammonia, through more complex but common chemicals such as formaldehyde and methyl and ethyl alcohols, to long atomic chains that make exotic organic molecules. We have little idea of how far the complexity extends, but there is strong evidence for an amino acid, glycine.

Though not themselves found, benzene molecules (which are in the shapes of rings) can link together into much larger observed structures called polycyclic

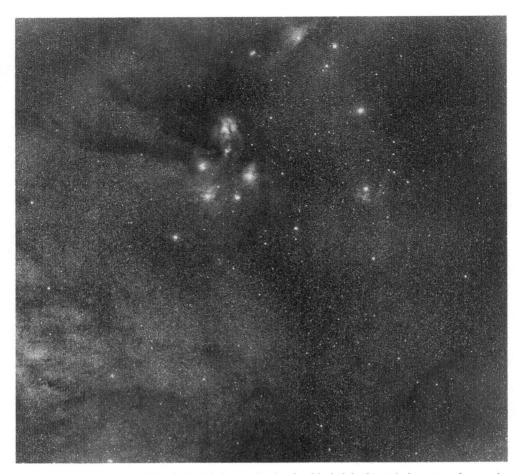

Figure 8.4. The Rho Ophiuchi giant molecular cloud is dark in the optical spectrum because its dust blocks the light of background stars (Rho Ophiuchi is buried in the bright nebulosity toward the top). In the radio spectrum, however, the dark cloud is bright with emissions from molecules, particularly carbon monoxide, which traces the much more abundant (but hard to observe) molecular hydrogen. [From the *Atlas of the Milky Way*, F. E. Ross and M. R. Calvert, University of Chicago Press, 1934. Copyright Part I 1934 by the University of Chicago. All rights reserved. Published June 1934.]

(because they are in cycles) aromatic (because they smell) hydrocarbons (made of carbon and hydrogen), or PAHs, of which there are several kinds. At the extreme, the PAHs have the characteristics of small particles, blurring the distinction between gas and dust. Furthermore, the clouds display several emissions and absorptions that are not yet identified and probably are indicative of much more complex stuff. A major problem in the study of the molecular clouds is that the most abundant species, molecular hydrogen, does not produce strong absorption or emission spectra. The clouds are therefore generally traced by powerful emissions from the much less abundant carbon monoxide or by other compounds.

The dust and the molecules in the clouds are yet more intimately interlocked, as without the dust the chemicals could not exist. Molecules tend to be fragile and must be protected to survive. The dust blocks energetic starlight that could break them apart, and drops the temperature to only a few tens of degrees kelvin above absolute zero. At such low temperatures, molecules move slowly and, as a result, damaging collisions are minimized.

Equally important, the dust acts as a necessary medium to get chemical reactions going in the first place. The temperatures and densities are so low that gaseous atomic hydrogen does not combine with any efficiency to make the dominant molecular hydrogen from which the other molecules grow. Instead, the reactions take place on the dust, the same dust that afterwards protects the molecules that it births. Hydrogen atoms stick to individual grains and bond into the molecular form on the grain surfaces. They are then kicked off. Ionization of the molecules by cosmic rays that penetrate the cloud subsequently initiates a plethora of other reactions, and we are off and running for the creation of a wonderful chemical mix. The grains in turn become highly modified because heavier atoms and molecules stick to their surfaces in the form of chemical ices.

This chain of reasoning shows that new young stars are related to dusty molecular clouds (as was the Sun so long ago). If so, where are their predecessors, the "protostars?" Can we find stars caught in the act of forming and can we understand how they form? Can we "watch" the Sun being born? Indeed we can.

## Developing stars

Once again we use the Sun as a benchmark, and search for its younger versions, for stars that look something like the Sun but that are allied with the same dark clouds that we know contain, or could contain, obviously young stars like those of classes O and B. Our planetary system, which forms a disk around the Sun, provides a powerful clue. Over 200 years ago, the philosopher and scientist Immanuel Kant suggested that the planets were born from a spinning fluid disk, what we now would call the "solar nebula," with the controlling Sun at its center. Like the Sun, the planetary system should have had a predecessor. We therefore look for young stars that not only are associated with dust but that also display some sort of evidence for surrounding disks.

And there they are, swarming amidst the nearby dark clouds of Ophiuchus, Scorpius, Taurus, Auriga, Orion, and other constellations of the Milky Way. The variable star T Tauri typically shines at about apparent visual magnitude 11, but can erratically become as bright as tenth or as faint as 14th. It is the prototype of the "T Tauri stars," which clump into loose "T associations," and that are associated with – even buried in – dusty molecular clouds. Like the OB associations, T associations are gravitationally unbounded, that is, the stars are moving apart, a sure signal that they are young.

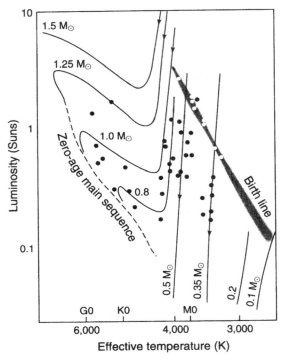

Figure 8.5. The T Tauri stars occupy a region in the HR diagram up and to the right of the main sequence. The forming stars have fired up their natural deuterium, become visible at the "birth line," and thereafter evolve toward the main sequence, their final positions depending on their masses. $M_\odot$ is the mass of the Sun. [S. W. Stahler, with evolutionary tracks by I. Iben Jr. (Art from *Astronomy!: A Brief Edition*, J. B. Kaler, © 1997, used by permission of Addison Wesley Educational Publishers Inc.)]

T Tauri stars at first look something like the Sun, typically exhibiting spectral classes of F, G, and K with surface temperatures of 4000–8000 K. There are, however, major differences. The T Tauri stars' luminosities place them well above (or to the right of) the main sequence. They at first look like they might be subgiants evolving off the main sequence to become giants, but they are in the wrong place on the HR diagram. Moreover, like solar-type stars, real subgiants are not related to dust clouds. The T Tauri stars, however, have intimate relationships with dust that go far beyond mere proximity. They are also powerful infrared emitters, producing vastly more of this longer-wavelength radiation than does the Sun. There is no readily available natural source of this radiation other than heated particulates, that is, heated dust. From the intensity of the radiation and a measure of temperature provided by the dust's infrared spectrum, we find that there is so much dust that the star should be invisible, that it should be buried like the windiest of Mira variables. For the stars to be visible, the surrounding dust cannot be spherically distributed. Distribution in a disk, however, works just fine. We seem to have found our predecessors! Instead of T Tauri stars moving off the main sequence like the subgiants and giants, they are moving onto it, and are the earlier stages of the stars of the lower main sequence, from class B (where they exhibit somewhat different behavior) down even to class M.

More powerful evidence of extreme youth is provided by the T Tauri stars' spectra, which contain strong lines of the element lithium. This fragile element is easily destroyed by thermonuclear processes when convection circulates stellar surface gases into deep, hot layers. Older stars like the Sun have little or no lithium; but T Tauri and its kind have a full complement equal in relative proportion to that of the interstellar medium. They have had no chance to destroy it, so they must indeed be young.

To grow, stars must accrete matter. In addition to their infrared signatures, T Tauri stars also emit strongly in the ultraviolet. Such radiation is consistent with hot matter raining violently downward onto the new star from its surrounding disk, now properly an accretion disk since it is the source of accreting mass. However, T Tauri stars confusingly display P Cygni lines (emissions with violet-shifted flanking absorptions) that show them to be *losing* mass. Conversely, some stars display the reversal of the P Cygni phenomenon, the absorption line placed on the long-wavelength side of the emission, showing that thick gas is indeed sometimes falling inward rather than outward. In fact, the stars seem to be both accreting and ejecting mass at the same time. Accretion apparently just wins the battle. Statistically, all this infrared, ultraviolet, and spectral activity is quieter for stars closer to the main sequence, supporting the idea that the stars are in fact developing into ordinary dwarfs.

The ejected matter goes back into the interstellar medium, where it has a potent effect. Many T Tauri stars are related to strange clumpy clouds of gas independently discovered 40 years ago by George Herbig of Lick Observatory and Guillermo Haro of the University of Mexico. Herbig–Haro objects, which have no readily apparent source of illumination, tend to come in pairs several light-years apart. In the middle, we commonly find a T Tauri or related star, from which emerges opposing jets that point toward, or even into, the Herbig–Haro objects. Where a jet is not visible, it is rather clearly hidden by local dust. Given that T Tauri stars possess disks, it is likely that the mass being lost is moving outward in a bipolar flow along the rotation axis

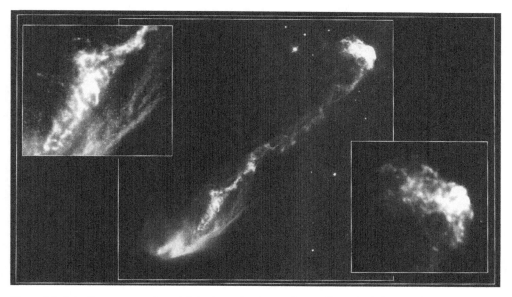

Figure 8.6. A jet from a star at lower left in the main picture plows a "bow shock," which in the lower right inset is seen as a classic Herbig–Haro object. The upper left inset shows detail in the turbulent jet. [J. Morse *et al.*, STScI, and NASA.]

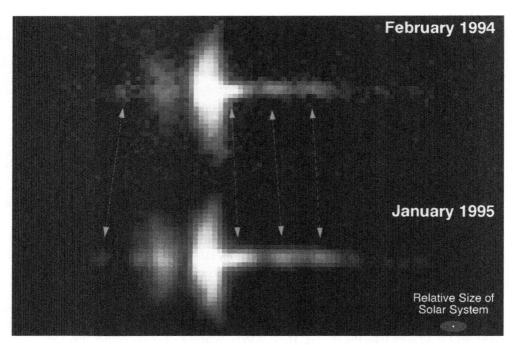

Figure 8.7. Closeups of a young star taken with the Hubble Space Telescope show a disk set perpendicular to emerging jets that down the line create Herbig–Haro objects. The star, invisible in the dusty disk, illuminates the disk's edges. The arrows show the outward movement of blobs of matter along the jet, speeding along at 200 km/s. The relative size of the planetary system is illustrated at lower right. [C. Burrows *et al.*, STScI, ESA, and NASA.]

of the star–disk combination. We see the same kind of phenomenon at stellar birth as we do as stellar death, bipolar flows also playing powerful roles in the structures of planetary nebulae and in the ejecta surrounding luminous blue variables.

The Herbig–Haro objects are caused by shock waves from flows that are hitting the interstellar medium. They even have a characteristic "U" shape as the flows plow ahead through the ambient gas. The flows are complex, and consist of ionized cores within cooler molecular flows, and may be expelled by magnetic fields that are generated by the rotation of the disk around the developing star. In a few cases, we can actually see the disks themselves with both radio and optical observations. Several are found within the Orion Nebula, which, with its O stars, is a hotbed of star formation.

Much of the accretion of matter for the new stars seems to take place in brief, violent episodes. The accretion disks from which mass falls to the stars are unstable. A disk can dump so much energy onto a T Tauri star that it will suddenly brighten by many magnitudes, far above the level of the usual variability. The classic case is FU Orionis, which went from fainter than 16th apparent visual magnitude to 10th magnitude in under a year in 1937 and is still up there. Several of these "FU Orionis stars" are known, including V1016 Cygni, which brightened by 6 magnitudes in 1969. To achieve this kind of brilliance requires a mass-infall rate of an astonishing hundredth of a solar mass per year. Over the course of 30 or 40 years, FU Orionis and its kind must acquire a significant fraction of a solar mass from the surrounding pre-stellar nebula, during which time the mass-gain rate far exceeds the mass-loss rate.

The O and hotter B stars do not form in quite the same way as those on the lower parts of the main sequence. They develop so quickly that they arrive at the main sequence while still accreting matter, still surrounded by thick accretion clouds and by diffuse nebulae. In this way they point to the locations of the origins of the lesser stars of the main sequence. The Orion Nebula is filled with young stars packed to a density of well over 100 per cubic light-year, vastly greater than the number we find near the Sun.

## Suns and planets

For the final connection, we must relate the T Tauri stars to the main sequence and connect the disks that circulate around T Tauri stars to our own planetary system. Here at home, the disk consists of many large bodies; out there, among the stars, we see disks of dusty gas. Theory, combined with computer simulation, does the job.

T Tauri stars are contracting under the action of gravity, their interiors heating. Mature stars like the Sun run off hydrogen fusion, off the proton–proton chain. For the chain to operate, the interior temperature must be above the flash point of around 7 million kelvin. As stars develop from accretion of matter, they are at first too cool for the chain to operate and must shine simply from the conversion of gravitational energy into heat. Deuterium, however, fuses at a lower temperature than does

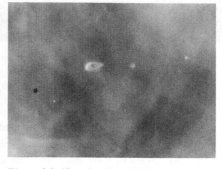

Figure 8.9. (See also Plate XXIII.) Associated with the Orion Nebula are numerous disks surrounding young stars, most lit by the bright Trapezium stars that power the nebula. One dark disk to the left is seen in relief against the bright gas. [C. R. O'Dell, Rice University, STScI, and NASA.]

Figure 8.8. (See also Plate XXII.) FU Orionis, a massively brightened T Tauri star, is surrounded by a large dusty cloud. [T. Nakajima and D. A. Golimowski, Johns Hopkins University and Palomar Observatory, California Institute of Technology.]

Figure 8.10. A tip of a huge column of dust within the Eagle Nebula in Serpens is being evaporated away. The small "finger tips" seem to be stars emerging from the deep gloom of the dust. [J. Hesterand P. Cowan, Arizona State University, the STScI, and NASA.]

normal hydrogen. As contraction proceeds, the temperature first reaches the fusion point of the star's natural deuterium (which it got from the interstellar medium), an event that seems to mark the beginning of much of T Tauri behavior. A star is first in a state of convection in which fresh deuterium is brought from the outer envelope into the fusion zone. It then dims at roughly constant temperature, dropping downward in the HR diagram from the "birth line" at which it first becomes placeable on the HR diagram. As a young star settles toward the main sequence, the disk begins to empty of its matter. Moreover, the star's wind can blow some of the interior disk matter away, so the accretion rate dramatically drops and the T Tauri behavior begins to die off. As deuterium fusion and convection die away, the star (at least one like the Sun) swings to the left on the HR diagram, now heating at its surface at about constant luminosity. When the temperature of the interior hits the fusion point of hydrogen, the star quickly enters the main sequence, and a new sun is born, one that will live its life quietly as long as there is any hydrogen left in the core at all.

Some of these new and developing stars have cleared their birth clouds, allowing them to be seen. Vast numbers of others throng the interiors of the dusty clouds associated with diffuse nebulae, where they are observed with infrared telescopes. Nearby O stars can evaporate the dust and molecular gas, slowly revealing them. Others simply escape, all to roam the Galaxy for their main sequence lifetimes, someday to produce a planetary nebula that will announce that another star is about to die.

Whatever the scenario, the final result is a new, or almost new, star with a residual disk, one that no longer supplies new matter. The disk, however, is far from peaceful. It is filled with dust, the grains moving in orbit around their star. Individual grains bump into one another and stick, the dust slowly accumulating into larger particles. Bigger particles enlarge at the expense of smaller ones, and at some point, gravity becomes a consideration, allowing even greater growth into primitive "planetesimals" some hundreds of kilometers across that are conglomerates of interstellar dust and dust that has freshly condensed within the solar nebula.

The chemical compositions of these small bodies depend on distance from the star, now our own Sun. In the interior, close to the Sun, volatile compounds like water cannot condense and stay in the vapor state. Only beyond about 4 or 5 AU is the temperature cool enough to allow the accretion of solid water. The result is that the bodies close to the Sun are rocky–metallic, made of elements and compounds that freeze to the solid state only at high temperature. Those far from the Sun, however, can incorporate volatiles like water, methane, ammonia, alcohols, and various hydrocarbons.

The inner planetesimals look like common asteroids, the rocky and metallic bodies that swarm in the spaces mostly between Mars and Jupiter. Asteroids range in size from about 1000 km across for the largest, Ceres, down to motes of dust. Small (and sometimes large) asteroids continually land on Earth as meteorites, allowing us to study them up close, where we see they consist of rock somewhat similar to that in the Earth, or of nearly pure metal, iron and nickel.

The outer planetesimals are still seen as comets, dirty iceballs typically a few tens of kilometers across that have been sent into long elliptical orbits that occasionally bring them close to the

Figure 8.11. The rocky asteroid Ida, 50 km across, was imaged by the Galileo spacecraft on its way to Jupiter. Smaller asteroids hit the Earth as meteorites. Ida is heavily pock-marked by collisions with other asteroids. [Galileo image, NASA/JPL.]

Figure 8.12. Halley's Comet, seen in its 1986 appearance, is an ancient remnant of the icy primitive bodies that made the outer planets. Its streaming tail, ejected from a nucleus only about 10 km across, is made of gas, as well as bits of dust and rock that may some day be seen as meteors. [Mt Laguna Observatory, University of Illinois, San Diego State University.]

Sun. Solar heat evaporates the ices and ionizes the resulting complex gases. The solar wind coupled to the Sun's streaming magnetic field then blows a comet's glowing gas backward into a tail that can be up to an astronomical unit long. The rocks and dust released from the comet's small nucleus fly away to be seen as a second dusty tail that is blown back and illuminated by sunlight. Bits of the fragile fluff from generations of comets rain upon the Earth to streak through the sky as common meteors (the particles generating bright shock waves), the pieces so small that they never hit the ground.

The inner primitive rocky planetesimals collided, the bigger winning out until they created the terrestrial planets: Mercury, Venus, Earth, and Mars. The violence of the impacts at least partially melted the early planets. The heavy stuff – iron and nickel – sank to the center, and the light stuff floated to the outside, leaving us with a planet with an iron core (which produces our magnetic field) surrounded by a rocky silicate mantle. On top floats the lighter cool crust.

Jupiter and Saturn were in a region so chilly that they could accumulate hydrogen and helium gas from the primitive solar nebula and could grow very large, and now have deep liquid-hydrogen layers surrounding rocky cores. Uranus and Neptune formed in a similar way, but in a region in the outer disk in which the raw material was sparser, so they came out smaller and quite different from Jupiter and Saturn.

The small bodies that now constitute the asteroid belt were so disturbed by giant Jupiter that they could not accumulate to form a planet. (For much the same reason, Mars came out smaller than Earth.) Subsequent violent collisions broke the asteroids back down to yet smaller bodies. The ones that had already differentiated to iron cores produced the iron meteorites we see today, their stripped-away outer layers a good portion of the rocky meteorites. Leftover comets beyond Neptune, where they were too few in number to accumulate into a planet, are still there, and are being revealed as the bodies of the "Kuiper belt" (honoring Gerard Kuiper), of which Pluto seems to be the chief member. (Though most likely a part of the Kuiper belt, Pluto is not considered a comet; it is too large and evolved, having developed from smaller cometary bodies.) Other leftover comets were hurled by the giant planets into a huge spherical volume centered on the Sun called the Oort comet cloud (after Jan Oort), which may contain a trillion of them. These primitive iceballs now leak back to us from both the Kuiper belt and Oort cloud to give us the short- and long-period comets of today.

The amount of the debris was much greater in the early days of the Solar System. The record of these leftovers is still seen in their heavy bombardment of the larger bodies of the Solar System, epitomized by the amazing number of ancient

Figure 8.13. (See also Plate XXIV.) The biggest of the terrestrial planets, Earth, is contrasted with the Moon. The other terrestrials, Mercury, Venus and Mars, are similarly structured, and consist largely of iron and rock. The apparent proximity of the Moon to the Earth is the result of foreshortening. [Galileo image, NASA/JPL.]

Figure 8.14. (See also Plate XXV.) Jupiter, king of the planets, incorporates light stuff, and is made mostly of hydrogen and helium. Its outer large satellites are largely water ice. [Voyager image, NASA/JPL.]

185

Figure 8.15. The Moon's craters preserve a record of immense bombardment by early, primitive planetesimals, some of which were very large, as revealed by the huge circular, lava-filled basin at top right, Mare Imbrium. [UCO/Lick Observatory image.]

impact craters on the Moon, a body so small that no geologic processes as we know them – erosion and weathering – have ever worn them away. The only thing that destroys craters is further bombardment, evidenced by the huge impact basins that form the dark lunar maria.

These ideas, which are still in a state of flux and have significant gaps, nevertheless provide a clear route from the disks of the T Tauri stars to our own present state, showing us both how our Sun formed and how our own home was created from the first primitive developing stars. If this picture is correct, planetary systems should be all around us, and we are in fact finding them. There is clear evidence for warm dusty disks around many mature stars. Vega, Fomalhaut, and other stars show the characteristic infrared signature. The most famous example is Beta Pictoris, which has a thick, visible disk seen edge-on extending 400 AU from the star in each direction. Several other similar disks have been found as well. Spectra of the Beta Pictoris disk show evidence for silicate and carbon dust, and observations with the Hubble Space Telescope indicate clumps of matter falling toward the star. Are there planets within? Most astronomers feel strongly that there are.

Actual planets are now being revealed not by direct imaging (which is yet not possible) but by the reflexive motions caused in the parent stars by the gravity of the planets. The planet behaves like a low-mass companion in a double-star system, and the star swings back and forth along the line of sight. From the periodic Doppler shifts, we can determine the size of the planetary orbit and even obtain an estimate of mass. A similar technique was applied to the "Doppler" shifts seen in the pulses from pulsars that not only revealed planets but showed that it seems almost impossible *not* to get a planet to form.

But we are really looking for more familiar bodies around normal stars. Several dozen such are now known, most about a Jupiter mass or larger ("earths" are as yet beyond detection). What we find is surprising, though perhaps it should not be. Extra-solar planetary systems were parochially assumed to be like ours, but they are not. Many of the first ones found have their "Jupiters" close to their parent stars

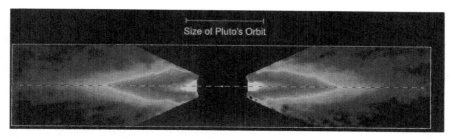

Figure 8.16. (See also Plate XXVI.) A dusty disk, 800 AU wide, surrounds the class A main sequence star Beta Pictoris. Evidence that the disk may contain planets includes the chemical composition of the dust, the warping of the disk, and an interior hole where planets may have been accumulated from the dust. (The "hole" seen here is produced by an occulting device that hides the bright star.) [C. Burrows and J. Krist, the STScI, ESA, and NASA.]

where our terrestrial planets are. Theory suggests that these large planets may have formed farther out and were brought in by friction against a thick residual disk that had yet to dissipate.

We have yet to see and study them, however. Are they really planets as we know them, are some brown dwarfs that are masquerading as planets, or are some of them something else yet? Many surprises await a growing science, though most of these bodies really are expected to be planets. If so, these systems must have come out of the spinning disks that once surrounded T Tauri – or similar – stars, in the same way in which our own planets were born so long ago.

## Back to the beginning

If the T Tauri (and more-massive) stars were created out of the dark dusty molecular clouds with which they are associated, we should be able to go back farther in time to still younger objects, perhaps all the way back to the very youngest ones created within the hearts of the dark clouds. What preceded the T Tauri stars? As a star develops, it clears away its surrounding birth cloud and eventually, like a T Tauri star, becomes visible to the eye, and ultimately, like the Sun, escapes from its birth cloud. At earlier stages of development, the forming protostars are still deeply buried and optically invisible. The infrared and radio regions of the spectrum now become especially crucial for the examination of stellar development, as these long waves easily penetrate the dusty shrouds. Within some clouds, radio telescopes detect long streamers of carbon monoxide radiating from a central source. Where there is CO there must also be molecular hydrogen and a powerful mass flow.

The radio spectra of different molecules are emitted under different conditions. Those of carbon monoxide show the location of the bulk of the gas, but ammonia lines are produced with strength where the gas is especially dense. Perpendicular to the CO (and $H_2$) flows, radio astronomers observe gas that radiates the ammonia lines, evidence for disks that will eventually surround visible T Tauri stars and, in a mature form, stars like the Sun.

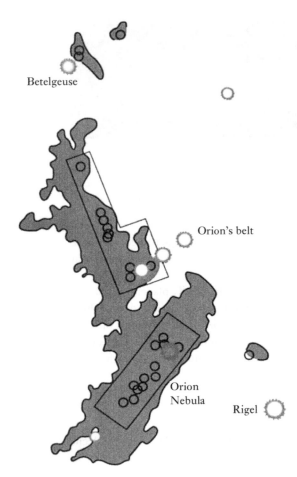

Betelgeuse

Orion's belt

Orion
Nebula

Rigel

Figure 8.17. New stars, marked by molecular flows (black circles), are coming into being within Orion's giant molecular clouds, which are outlined in gray against the familiar stars of the constellation. [R. J. Madalena *et al.*, rendered in *Cosmic Clouds* by J. B. Kaler © 1997 by Scientific American Library. Used with permission of W. H. Freeman and Company.]

We can go back even farther. Within the dark molecular clouds are knots of ammonia emission from "dense cores." They represent the very beginnings of the stars as they start their slow collapse from the molecular clouds of the interstellar medium. Here are our ultimate origins, the youngest "stars" of all. The Sun was almost certainly associated with a dense core five billion years ago, one that had all the raw materials that were the origins of everything we see around us.

To look further at beginnings, return to the O stars. They do not travel far before evolving to supergiants, perhaps producing expanding ring nebulae prior to exploding as supernovae. The powerful shock waves blast through the surrounding medium, and when they encounter a thick cloud compress it, causing the denser parts of it to begin the process of collapse. (Other compression mechanisms work too: the density enhancements caused by the Galaxy's spiral arms, cloud collisions, perhaps even winds from O stars.)

It would seem that all that is needed is to allow the collapse to proceed right through to the creation of a star. However, the dense core, or what will become the dense core, must be rotating, if only as a result of collisions with its neighbors. As a rotating body contracts, it has to spin faster, the result of the "conservation" of angular momentum. Swing a weight tied to a string around your head. The angular momentum is the mass of the weight times the length of the string times the weight's velocity. In the absence of an outside force, angular momentum must be conserved, that is, stay the same. Bring in the string to decrease the radius and the weight moves faster, a skater brings in her arms and becomes a blur.

When a body rotates, centrifugal force makes it bulge at its equator, the solid Earth 70 kilometers wider across its equator than it is through its poles. Saturn, mostly in the liquid state and spinning much faster, displays a ten percent difference

Figure 8.18. Galaxies are constituted of stars. But it takes an entire galaxy to *make* its stars. Here we see clumps of new stars that have formed within the spiral arms of the galaxy M101. [AURA/NOAO/NSF.]

from equator to pole. A blob of dusty gas contracting without interference from a dense core a fraction of a light-year across would, by the time it was closing in on the size of the Sun, be rotating so fast that it would tear itself apart and could not form a star. For stars to exist, something must help the contracting protostar lose its angular momentum. The search for the mechanism of star formation has largely been that for the means for the removal of spin.

Starlight cannot penetrate the dusty dense cores, leaving them filled with cold neutral atoms and molecules. However, the cores *can* be invaded by the Galaxy's enormously energetic cosmic rays, which ionize atoms and molecules to aid in the creation of the cores' complex chemistries. The rotating Galaxy generates a weak magnetic field that threads its way through all of interstellar space. The few ions in a dense core grab onto this magnetic field, which acts as a brake, and slowly but inexorably slows the rotation. As a core contracts, the ions migrate outward under the effect of the magnetism, and the field eventually loses its grip, but not until after much of the heavy work is done.

A core may still be spinning fast, but now other mechanisms can take over. It may indeed yet tear itself up. But instead of shattering, it separates into two blobs of gas that independently contract, creating a double star, spin energy going into orbital energy. If the spins are still too fast, each may yet again separate into doubles to yield a double double-star like Epsilon Lyrae. The residual rotation of a single protostar (and perhaps even the components of a double) coupled with the conservation of angular momentum will make the outer parts of the collapsing core flatten into a disk from which the growing star will accrete much of its matter. The disk will eject flows along its rotation axis, helping to slow the system even further, and we return to where we started: to a T Tauri star and its surrounding accretion disk, from which a new sun and perhaps a set of planets will someday develop.

We see in looking at the youngest stars that their formation is anything but isolated. To a remarkable degree, the Galaxy, star formation, and stellar evolution are tightly interlocked. We, our Earth and ourselves, are the creation not just of some local cloud, but of other stars and their ejecta. Through the action of its thronging population of aging stars and of its magnetic field, we are the children of the entire Galaxy.

# Chapter 9

# The oldest stars

Stars are members of the Galaxy, which is instrumental in their formation. Not only does the Galaxy contain the necessary raw material, but it also provides the means of compression of that material through its spiral arms and by the blast waves of high-mass supernovae. So the ultimate origins of the stars lie in the origin and evolution of the Galaxy itself. The exploration of young stars therefore contrarily requires the study of the oldest stars to find ancient remnants of Galactic formation, the clues to Galactic development, and the seeds that formed the stars that are being born today.

## Open clusters

We know the Sun to be middle-aged in terms of its potential main sequence lifetime, about 5 billion years old out of a total of 10 billion, at which point the interior fuel will be exhausted and the Sun will die, first as a red giant and then as a white dwarf. We also know that there are younger stars, typified by O and B stars both in and out of young clusters, and by T Tauri stars that flock within dark clouds. But are there stars older than the Sun? If so, where are they and what do they look like?

The "oldest stars" fall into a variety of categories that depend on where they are located. Examine the most populous part of the Galaxy, the disk that makes the Milky Way. Here we can look at groups of stars from which we can find ages, the open clusters. Young clusters are identified by the memberships of O and B stars, whereas older clusters are missing their upper main sequences. Now look to the extreme to find clusters that are missing as much of the main sequence as possible. It takes around a billion years for the coolest (lowest-mass) A dwarf to use its interior hydrogen and become a giant. So if we see a cluster with no A dwarfs, we know the cluster must be about one billion years old. If the F stars are missing too, and the main sequence stops in class G, the age must be closer to ten billion years. By matching theoretical predictions of stellar aging to the observed HR diagrams, specifically

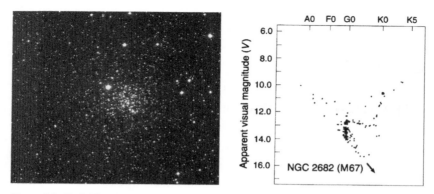

Figure 9.1. (*Left*) The oldest prominent open cluster is M67 in Cancer. Its HR diagram (*right*) lacks an upper main sequence. The position of the giant branch shows the cluster to be about 5 billion years old. [© National Geographic–Palomar Observatory Sky Survey, reproduced by permission of the California Institute of Technology; HR diagram adapted from *An Atlas of Open Cluster Color–Magnitude Diagrams*, by G. C. Hagen, Publ. of David Dunlap Observatory, Toronto, 1970. (Art from *Astronomy!*, J. B. Kaler, © 1994, used by permission of Addison Wesley Educational Publishers Inc.)]

to the point at which the giants join the main sequence, astronomers can be rather precise, identifying ages to within a billion years or so.

The open clusters that throng the Galaxy's disk provide a chronology. The disk must be at least as old as its oldest open clusters, so we look for the oldest we can find. The search is difficult, as the large majority of open clusters are not only not old, nor even middle-aged, but young. There is no reason to think that more clusters are being made now than were made in the past, so something must be destroying them. Open clusters are loose affairs. They do not contain many stars – perhaps a few hundred – and the mutual gravitational forces that bind them together are relatively low. When two stars in a cluster pass close to each other, they can gravitationally transfer energy: one may slow down, the other speed up. On occasion, one star will gain enough speed to exceed the cluster's escape velocity and will be ejected. The process is powerfully aided by tides raised by the Galaxy and induced by encounters with giant molecular clouds, which pull on, stretch, and disrupt the clusters. An open cluster therefore slowly evaporates.

Are there any old ones at all? Yes, but only dense, more tightly bonded ones that are also outside the Sun's Galactic orbit. Beyond 25,000 light-years from the Galaxy's center, where galactic tides are weaker and the amount of disturbing matter is much lower than it is toward the inside, clusters can survive the onslaught of gravity. The best-known older cluster is M67 in Cancer, easily found with a small telescope in northern springtime skies not far from the much younger Beehive cluster. Yet though it is among the older clusters known, evolutionary analysis shows it still to be only five billion or so years old, about the same age as the Sun.

So we must look to more obscure systems. About a dozen clusters rank older than M67, which at solar age is also something of a benchmark. The better known, those with common catalogue names, are NGC 188 in Cepheus and NGC 6791 in Lyra. Both are accessible with a decent amateur telescope, and both are estimated to be about eight billion years old, three billion more than the Sun. (When the Sun was born they were still losing their class F stars, which have since evolved into white dwarfs.) NGC 6791 may in fact approach 10 billion years, depending on whose study one adopts. The oldest known is an obscure open cluster in Auriga called "Berkeley 17," which seems to fall between 10 and 13 billion years old. When all are counted together, the greatest reliable ages seem to be around 10 billion years.

Our Galaxy's disk is therefore around 10 billion years old. Could there be some that would have been older had they not disrupted? If so, 10 billion years is only a lower limit. It seems unlikely, as the oldest clusters are still quite vigorous in appearance and hardly on their last legs. Ten billion years seems fairly secure – give or take a billion or so. Moreover, this number fits well with the age found from the dimmest white dwarfs. Our Sun came along when the Galactic disk's age was half what it is today.

## Into the halo

To go back farther, look into the vast, thin, outer halo that surrounds our Galaxy's disk. As Andromeda climbs above the eastern horizon on a northern autumn night, locate a favorite sight, the Andromeda Galaxy, M31. The most distant object the naked eye can see, it appears as an elongated fuzzy patch, and even a large amateur telescope shows little more, only an amorphous mass of stars. The outer parts of this galaxy were resolved into individual stars in 1923 by Edwin Hubble with the 100-inch telescope on Mt Wilson, finally revealing that M31 is much like our own Galaxy. Indeed he finally and securely demonstrated that our own Galaxy is just one of countless others that come in many different forms, some with spiral arms like ours and M31, others ellipsoidal without either arms and star formation, yet others ragged and without much form at all.

The inner portions of M31, however, defied resolution. During

Figure 9.2. NGC 6791 (*top*), with an age of 8 to 10 billion years, is one of the oldest of open clusters. The oldest known, Berkeley 17 (*bottom, center*), is barely distinguishable from the stars of Auriga's Milky Way. [© National Geographic–Palomar Observatory Sky Survey, reproduced by permission of the California Institute of Technology.]

193

Figure 9.3. The Andromeda galaxy, M31 (*arrow*), seen in the context of its constellation, is an easy sight in a dark northern autumn sky. Beta Andromedae is down and to the left of M31, while Gamma is at far left center. [Author's photograph.]

World War II, when the skies above Los Angeles went dark as a result of wartime blackouts, Walter Baade attacked the central bulge with a new kind of (black-and-white) photographic emulsion that was sensitive to red light. There, for the first time, he saw masses of red giants. As he moved his photographic investigation in to the disk, however, the color changed. The spiral arms were blue, filled with the O and B stars of the upper main sequence. That is the only place they appeared. Baade separated the two kinds of stars by calling those of the disk Population I and the those of the bulge Population II.

Most of the stars in the solar neighborhood, including the ones that make the friendly constellations, are of the Population I variety. As such, they have more or less circular orbits about the Galaxy, as does the Population I Sun. As a result, the motions of these stars relative to each other are not very great. From our solar point of view, the velocities – calculated from proper motions and radial velocities – are not very high, typically 20 to 30 kilometers per second. While that may seem huge – New York to Los Angeles in two minutes – compared to the sizes of stars and the distances between them, it is small. At 25 km/s it takes a star like the Sun 17 hours to move through its own diameter, and it would take 50,000 years to go from here just to the nearest star, Alpha Centauri. So the constellations change in a leisurely fashion.

Figure 9.4. A telescopic view of the Andromeda galaxy shows an easily resolvable outer disk and a nearly impenetrable dense central bulge (which is the part seen with the naked eye). At the bottom is a small elliptical galaxy companion (M32) that contains only old stars. [AURA/NOAO/NSF.]

However, the stars of the halo are in elliptical orbits, needed to take them out to the great distances at which we see them. Since the Galaxy's disk slices through the halo, we must have halo stars around us; there are just not as many as actually belong to the disk. Though at first glance a halo star looks like any other, its elliptical orbit, quite different from that of the Sun, ensures that the star must be moving at a high speed relative to us. Most halo stars, for example, are plunging through the disk while

195

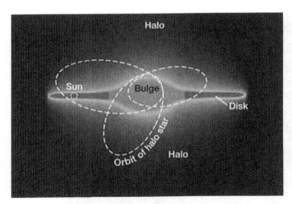

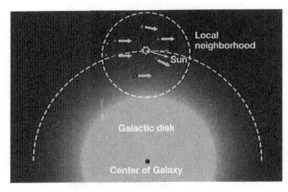

Figure 9.5. (*Lower*) The stars of the Population I disk have more or less circular orbits about the Galaxy's center, and therefore are not moving very fast relative to each other. (*Upper*) Those of the Population II halo move through the disk in elliptical orbits, so we zip past the local Population II stars at high velocity. The globular clusters are halo objects, and therefore also have high velocities. [From *Astronomy!: A Brief Edition*, J. B. Kaler, © 1997, used by permission of Addison Wesley Educational Publishers Inc.]

we sail past them in a near-perpendicular direction. Relative speeds are therefore typical of the Sun's actual orbital velocity of over 200 kilometers per second. Barnard's Star, the famed M dwarf with the highest proper motion, clips along at 140 km/s relative to the Sun, and Kapteyn's Star, a ninth magnitude M0 dwarf in the southern constellation Pictor, whizzes away from us at nearly 300 km/s. Both of these "high-velocity stars" must belong to the halo.

There are very few blue O or B or even white class A high-velocity stars among the high-velocity set. The HR diagram of the high-velocity stars effectively extends from about class G on down the main sequence and up into the red giants. When put together as a group, they would appear reddish, just like M31's bulge. The high-velocity stars thus have the original characteristic of Population II, showing that this Population extends into the enormous but faint halo. The halo is nearly all Population II.

## Globular clusters and Population II

Within the halo, we see much older stars. As Andromeda rises, look in the other direction to see Hercules descending the western sky, where you can see another fuzzy patch, M13. A large telescope turns it into a spectacular spherical mass of stars, a globular cluster. Globular clusters, very different from open clusters, are densely packed with stars. A rich globular, like M13 or Omega Centauri, has a radius of 100 light-years or so, and within that volume contains as many as a million stars. The clusters are strongly concentrated toward their centers. At the core of M15 in Pegasus, the Hubble Space Telescope finds 7000 stars within a region only 0.7 light-years across, a density a million times greater than that in our own local neighborhood.

Spectacular they may be, but there are not very many of them. Only about 150 are known. A few must be hidden behind the dense dust of the central Milky Way,

Figure 9.6. One of the great globular clusters of the northern hemisphere, M15, shows off its extraordinary central condensation. [WIYN/NOAO/NSF, S. Slavin, H. Cohn, P. Lugger, and B. Murphey.]

but the total count probably does not exceed 200. Nor are they all M13s or Omega Centauries. Many have far fewer stars and smaller radii. At their least they are only a third the size of the big ones and contain only a few thousand stars. Whatever their individual characteristics, however, the globular clusters as a group have a number of singular characteristics. Most importantly, they are not confined to the Galactic disk, but concentrate in our Galaxy's bulge in Sagittarius and then spread far into the halo as they move about the Galactic center on elliptical orbits. Globular clusters are emblematic of Population II.

Globulars also distinguish themselves from open clusters and Population I by their HR diagrams, which provide profound clues for understanding not just the clusters but the entire halo. Whereas open clusters display a large mix of stars, some with upper main sequences, others without, no globular – as befits halo stars – has an upper main sequence. The brilliant O and B stars are all gone, having long ago burned away. Instead, the globulars all display a feature that is missing from Population I, a distinctive "horizontal branch" that starts from the middle of the red-giant branch and extends leftward. The lack of upper main sequence stars – as also demonstrated by the lack of hot stars among the set of the high-velocity pack near us – implies that Population II as a class is old.

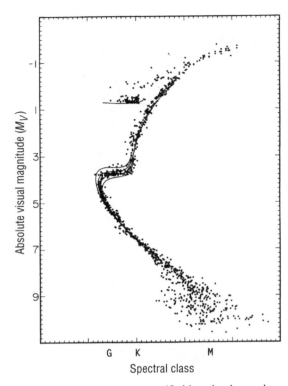

Absolute visual magnitude ($M_V$)

Spectral class

Figure 9.7. Globular clusters, typified here by the southern hemisphere's 47 Tucanae, have no upper main sequences. Instead, they have a distinctive horizontal branch that is missing in Population I. The scale of spectral classes along the bottom is only approximate, as the stars, low in metals, appear to be shifted somewhat to the left as compared with Population I stars. The curves give the loci of the stars expected for different ages, allowing the cluster's age to be determined. [J. D. Hesser *et al.* in the *Publications of the Astronomical Society of the Pacific*, 1987.]

Spectral analyses of the stars that belong to globular clusters show that relative to the Sun and to Population I in general, they are deficient in "metals." To an astronomer, the term implies anything heavier than hydrogen and helium. The deficiencies occupy an enormous range, from about a thousandth the iron content found in the Sun to approaching solar. Isolated halo stars show much the same characteristic, though some have even lower deficiencies. The lowered metal content increases the transparency of the stellar gases, shrinking the stars, and for a given mass increasing the temperatures (that is, for the same mass, Population II stars are the hotter), making the stars look bluer. The effect is exaggerated in that the lack of metal lines, which tend to congregate at shorter wavelengths, makes the stars look bluer yet. On the HR diagram such stars form their own line of "subdwarfs" seemingly below, but actually to the left of, the standard main sequence. The helium-burning giant stars of Population I concentrate into a clump about half-way up the giant branch. As a result of lowered metals, and in part because of small differences in mass, the helium-burners in globular clusters spread out to the left of the giants, creating the distinctive horizontal branch.

General appearances aside, globulars are not a homogeneous lot. Not only are some large and some small, but there are also at least two spatial distributions. The globulars with the least metals are widely spread into the outer depths of the Galaxy's halo, to distances of over 100,000 light-years, double the size generally given for the radius of the disk. The higher-metal globulars, however, are arranged in a thick disk, showing us a general, though vague, increase of metal content from the outer halo down into the thin Population I disk itself.

Lower metal content also makes spectral classification problematic. Classes are determined by the strengths of a variety of absorption lines. As temperature drops, Population I stars contain more metal absorption lines. Spectra of A stars for example, at first glance appear quite simple, whereas those of class G are filled with strong calcium, sodium, and other lines. If you remove metals from a star and do not take the altered composition into account, you will classify the star as hotter than it deserves. As a result, astronomers commonly drop spectral classes and use a "color index," which is responsive to temperature, instead.

If we use Population I spectral classes, the point where globular-cluster giants join the main sequence seems to be at a hotter class than it does for the oldest open clusters, and globulars seem to be the younger. However, when the metal deficiency is taken into account, the ages of the globulars are found to range from about that of the oldest possible open cluster to considerably older, some 13 billion years, perhaps even older. (The upper limit is vigorously argued.) All the evidence shows that Population II had to come first, before Population I; the halo must be older than the disk.

## Evolution of the Galaxy

The metal deficiency of Population II combined with theories of stellar evolution provide the means with which to connect the two populations and to find a simple scenario for the evolution of the Galaxy. We have looked at individual extreme stars that exemplify processes of stellar evolution. Now put the processes together and summarize. All stars lose mass, even the Sun through the solar wind, though at the paltry rate of $10^{-13}$ of itself a year. As stars age and turn into giants, the higher luminosity and lower surface gravities combine to allow much higher mass-loss rates. By the time a typical A, F, or G main sequence star becomes a Mira variable it is so huge, and the internal processes so active, that its wind can blow at a rate a hundred million times greater than that from the Sun, as much as a hundred thousandth of a solar mass a year. The O and B stars of the upper main sequence, with their initially greater luminosities, generate even fiercer winds as they evolve into supergiants.

The stars are in these states for tens if not hundreds of thousands of years, so as they evolve they lose a good portion of themselves back into space. For lower-mass stars, such mass loss is represented by the planetary nebulae, in which hot dying stars – essentially the old nuclear-burning cores – become surrounded by shells of expanding, now-illuminated matter. The Sun will about cut itself in half. A 10 solar-mass star will become a white dwarf near the 1.4 solar-mass Chandrasekhar limit, and lose 85% of itself. High-mass stars will lose even more, their mass loss represented by ring nebulae and ultimately by the powerful blasts of core-collapse supernovae that leave behind neutron stars and black holes. Stars therefore provide enormous amounts of matter for the clouds of interstellar space from which new stars can be born; stars are the ultimate recycling engines.

Stellar chemical production ties it all together. Stars shine as a result of the conversion of light elements into heavy elements by the various processes of thermonuclear fusion. Hydrogen first becomes helium, then helium becomes carbon and oxygen. Convection currents within mass-losing giants sweep some of the by-products of thermonuclear fusion into the stars' outer layers. The stars subsequently pump vast amounts of freshly made carbon, nitrogen, even helium, into their winds and planetary nebulae, along with the by-products of slow neutron capture (the s-process) that makes elements up to bismuth.

Massive stars carry the process much farther. Within the core of a supergiant, the carbon and oxygen created from the original hydrogen can fuse all the way to iron. The iron core, unable to support itself, collapses to create a supernova. The by-products of energy generation are exploded outward, as are those of explosive nuclear burning that makes iron and those of rapid neutron capture (the r-process) that can carry chemical formation beyond uranium. Type Ia (white dwarf) super-novae, those in which a main sequence star dumps too much matter onto a white dwarf close to the Chandrasekhar limit, causing it to collapse and explode (other scenarios are also possible), add to the inventory, and provide much if not even most of the iron. Supernovae can make everything! All this stuff, from both supernovae and evolving giant stars, gets added to the dusty gases of interstellar space and ultimately finds its way into new stars.

The processes of stellar evolution in fact are the *only* sources we have that can create elements heavier than boron, element number five in the chemist's periodic table. (Lithium, beryllium, and boron are made mostly in interstellar space by collisions of heavy atoms with cosmic rays; hydrogen, helium, and lithium were created in the Big Bang that began the Universe.) Any given heavy element is produced in varying quantities from the two chief sources, giants and supernovae. Carbon, strontium, zirconium, barium and others are made almost exclusively by giants; silver, most rare earths, platinum, gold, and uranium and more by supergiants and super-novae; others like rubidium, cadmium, tin, and tungsten by a combination.

As a result of this steady contamination, the interstellar medium should become progressively enriched in metals as the Galaxy ages. On the average, the newer the star, the more metals it should have. The Population I disk is metal-rich, the Population II halo metal-poor. The HR diagrams of the different sets of clusters have already clearly demonstrated the halo is older than the disk, that the halo came first. The chemical composition differences coupled with stellar evolution studies take us further, proving conclusively that the disk descended from the halo, that the characteristics of Population I came directly from the actions of Population II, and moreover showing why. The halo must have seeded the interstellar medium with heavy elements that gave later generations of stars their heavy elements.

The distribution of the two populations provides a profound clue as to how the Galaxy developed. The simplest possible – and classic – picture is that the Galaxy began as a great, nearly spherical cloud of matter that was contracting under the force of its own gravity. For reasons that we do not understand, the cloud fragmented into large clumps that then fragmented into stars: the first globular clusters. Since the cloud was contracting, these ancient globulars had a natural velocity toward the Galaxy's center that placed them into the elliptical orbits we see today. Individual halo stars may have been born as well, or the globulars may have partially disrupted to populate the halo, or both.

As the cloud squeezed down, it spun faster, and flattened into a disk, an action of the conservation of angular momentum, the same process that produces the disks from which descend a star's planets. Younger stars, those born in the disk, no longer had injection velocities toward the center, and wound up in circular orbits instead, paths characteristic of a spinning disk, the same reason the planets have circular orbits around the Sun. At the same time, the collapsing halo and, ultimately, the disk became enriched in heavy elements and metals as a result of continuous stellar evolution. The globulars that have more metals occupy a thick disk; metal-rich Population I occupies the thin Population I disk.

The real picture is far from being this neat, as there is anything but the expected clear and continuous change between chemical abundance, Galactic orbit, and age. Such correlations are seen only in the broadest view when comparing the gross properties of the halo with those of the disk. The thick disk of relatively metal-rich globular clusters, for example, does not seem to be particularly older than the more-widely distributed metal-poor collection.

Instead, there are growing feelings that much of the chemical evolution of the Galaxy took place on a local scale, with local contamination by supernovae and giant stars. The new elements were not distributed Galaxy-wide, and the development of a particular region depended on the local rates of star formation and therefore on the rates at which supernovae exploded, which to some degree were chaotic and accidental.

Equally important, our Galaxy almost certainly did not develop in isolation from others. We are surrounded by a host of small galaxies, from the Magellanic Clouds to tiny dwarfs not much more populous than the great globular clusters. Though many galaxies exist in isolated splendor, most are arranged in clusters. Like star clusters, some are rich (the Virgo and Coma Berenices clusters containing thousands of members), others poor. We live in one of the latter, our own "Local Group" consisting of about three-dozen galaxies dominated by ourselves and M31.

When we look deep into the cosmos within these galaxy clusters, where the individual units are relatively close together, we see awesome violence as galaxies crash into each other. The collisions do not involve collisions of stars; stars are so far apart that galaxies can actually pass through each other. But the gases of interstellar space *can* collide, causing compression and enormously enhancing star formation and therefore supernova production. Such events also raise huge tidal streamers in which stars are vigorously born. Under the right circumstances, the galaxies can even combine.

Large galaxies may therefore grow not just from a contracting blob, but also from mergers. The newly discovered "Sagittarius Dwarf" is falling into, and in fact is within, our own Galaxy, carrying along with it the well-known globular cluster M54. Other small galaxies near us may have come from tidal streamers raised by earlier disruptions of our Galaxy; after enough time they may combine again with another galaxy or perhaps come back home. The confused and chaotic correlations between orbit, age, and chemical composition may in part be due to other galaxies

Figure 9.8. A cluster of galaxies in Coma Berenices, 300 million light-years away, contains hundreds of members and a variety of kinds. [AURA/NOAO/NSF.]

that were in different states of evolution integrating themselves with ours. Such collisions can also enhance star formation and chemical return in one part of our Galaxy at the expense of other parts. Moreover, astronomers find large thin clouds of interstellar gas falling into our Galaxy's disk. Some of it may have come from the disk's supernovae that spray huge fountains into the halo, but some may also be more-or-less pristine stuff with few metals, and therefore may have the effect of *diluting* the disk's metal content. It is going to take an immense observational and theoretical effort to sort it all out.

Equally disconcerting is the existence of vast amounts of "dark matter" of an unknown nature. Stars in our Galaxy orbit in response to the mass interior to their orbits. Orbital velocities trace out the distribution of mass; the more of it, the faster the speeds. Studies consistently reveal that the outer parts of our Galaxy are orbiting much faster than they would on the basis the mass tied up in observed stars and nebulae (including dark clouds). Motions of galaxies within clusters reveal the same phenomenon, gravity supplied by great amounts of "dark matter." An observed Galaxy may represent only a few percent of all that is actually there. How dark matter factors in to the evolution of a Galaxy is obscure at best.

## Toward the beginning

Our Galaxy is not isolated from others; nor is it isolated from the Universe at large. To understand its current nature, including the natures of its stars, we have to look at the biggest possible picture. The search for the oldest stars takes us back to the beginning of time and space itself, to the event that seems to have developed the Universe we see today.

The key item is that the Universe is expanding. Within our Local Group, galaxies are milling about as a result of gravitational interactions, some now coming toward us, some going away from us with speeds told by the Doppler shift. The Andromeda galaxy is approaching at almost 300 kilometers per second, the Small Magellanic Cloud, one of our larger companions, receding at 170. (At that rate we would not have to worry about a collision with Andromeda for billions of years, and may never, since we do not know its direction through space). But once we get beyond the Local Group, the spectrum lines of all galaxies are shifted to longer wavelengths, to the red; they are therefore moving away from us.

Once we measure the galaxies' distances, we see (as first did Edwin Hubble in 1929) that they are moving at speeds that are directly proportional to their distances. A galaxy twice as far from us as another is receding at twice the speed, one three times thrice the speed. Although it seems that we are at the center of some sort of expansion, since speed is directly related to distance, all galaxies (actually clusters of galaxies) are moving farther away from all other galaxies, and everyone (is there anyone?) sees the same thing.

Though it looks like the Universe evolved out of some kind of explosion, the expansion we see is actually conceived as an expansion not of galaxies into space, but of space itself. The galaxies are not moving through space away from us, but are caught into the web of a space that is itself expanding whether there are galaxies in it or not. The expanding Universe is seen only on a large scale, however. A galaxy does not expand along with the Universe, nor does a cluster of galaxies, because gravity holds them together. It is only over large distances, within which gravity is weakened, that the expansion can take over and make distances between things increase.

If the recession is caused by an expansion of space, then the red shifts (a singular concept reduced to "redshifts") are not caused by the Doppler effect, but by one in which the rules that relate spectral shifts to velocities are expressible only by Einstein's relativity. Matter and energy are part of space and are caught within it. A distant galaxy sends a photon toward us at the speed of light. Though the photon has no mass, it is still affected by its spatial environment and, just like a galaxy, is caught in the web of space. During its travel-time, measured in millions, even billions of years, space expands, and so must the photon that is trapped within it. By the

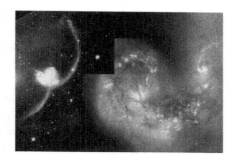

Figure 9.9. (See also Plate XXVII.) The "Ring-Tail galaxy" in Corvus consists of two colliding galaxies, NGC 4038 and NGC 4039, that send out long tidal streamers (*left*). A close-up of the interior taken by the Hubble Space Telescope (*right*) reveals hosts of new stars forming within the maelstrom of the collision. [B. Whitmore, STScI, and NASA.]

Figure 9.10. The Sextans dwarf is a small galaxy in the Local Group that exhibits the signs of star formation. Is it a fragment of a collision? Will it merge with another galaxy and help provide a confusing mixture of evolutionary properties for that galaxy's future astronomers? [Palomar Observatory, Cailfornia Institute of Technology.]

time the photon arrives here, it has a longer wavelength. The greater the duration of its flight, the more stretched and the redder it will become, neatly explaining the redshift relationship we see.

On top of the expansive redshift there will be normal Doppler shifts caused by the influence of gravity. Gravity also has the overall effect of slowing the expansion rate over time (unless there are other forces, like the energy of the vacuum itself, that can overcome the slowing or make it expand even faster). The rules that relate redshift to velocity depend on the exact "shape" of the Universe, on how space is distorted by gravity through the laws of Einstein's relativity.

An explanation of the observations is offered by the theory of the Big Bang. The mass of the Universe we see was at one time in the form of energy (Einstein showed we can convert back and forth). Crushed into a smaller volume, it was compressed into a state of enormously high density ($10^{100}$ grams per cubic centimeter) and temperature ($10^{31}$ kelvin); at a time before that we lose our way. Pressures within this dense system forced it to expand, but how it was created in the first place, no one knows.

Astronomers can, within limits, determine when this event happened. To find the time it took to separate another galaxy (rather its mass) from ourselves, divide its distance by its velocity. Because distance is directly proportional to velocity, the result – the age of the Universe – will be the same for all galaxies. The age is thus the inverse of the expansion rate, which is found by dividing any galaxy's velocity by its distance, and which is called the "Hubble constant" in honor of one of the founders of the field.

The history of twentieth-century cosmology – the study of the cosmos at large – is intimately linked to the measurement of the Hubble constant, which has been hotly contended. Numerous sophisticated studies, including those with the Hubble Space Telescope (using Cepheid variables and Type Ia supernovae, which are both reliable distance indicators), seem to be zeroing in on about 20 kilometers per second per million light-years. Inverting that number to obtain the age of the Universe gives 15 billion years, remarkably similar to the ages of the oldest globular clusters that define the age of our Galaxy and that are derived from completely different principles. We seem to be doing something right. The similarity is powerful evidence that the theory of the Big Bang is correct.

Not that all is perfect. The above age is correct only if there has been no deceleration of the expansion caused by gravity. But there can be no deceleration only if there is no gravity and therefore no matter. Such is hardly the case, as we see matter all around us, so the true age must be less than that found from the simple inversion of the Hubble constant. By how much we do not know. The simplest theory says that the Universe has no distortion (curvature) of space and that it will coast to a halt only after an infinite period of time. That is, there is just enough mass and gravity to bring the expansion to zero, but there is not enough ever to make it collapse. If that is the case, the true age is two-thirds that of the "no-matter" age, and therefore an age of ten billion years is closer to the mark. Thus the Universe would seem to be younger than its oldest globular clusters, so either the simple theory is wrong, the theory of stellar aging is at least partially in error, the distances of globulars or galaxies need some fine tuning, other forces are involved, or some combination is responsible. Indeed most current data suggest that the Universe does *not* have enough matter ever to make it stop, and therefore the actual age is greater than the two-thirds fraction and closer to that of the globulars. There may even be an expansive force (Einstein's "cosmological force") that may counter gravity, one that comes from the energy of the vacuum itself, which will slow or halt the expansion.

Ignoring the very earliest stages of the Big Bang and the means of the expansion, theory shows that within about a millionth of a second the Universe had cooled to the point where the energy would have frozen into protons, neutrons, and electrons. In stars, protons collide to make deuterium. Under Big Bang conditions, deuterium can be made by the combination of protons and neutrons. At high early energies, the colliding particles could not stick to form heavier elements, as collisions would break any new atoms apart. But as the temperature dropped to about a hundred million degrees kelvin (similar to that in an evolved star), about three minutes after the initiation of the Big Bang, the protons and neutrons formed deuterium, further reactions making helium and lithium. But the temperature was declining so fast that nothing more could be made. The result is that we were left with 92 percent so hydrogen nuclei, 8 percent helium, and a small amount of deuterium and lithium. And when we look at interstellar matter and at the oldest stars, that is pretty much what we see. Even the isotopic ratios of helium come out about right. Big Bang theory works! The higher helium abundances we observe within new stars and within the Galaxy's disk are the result of 10 to 15 billion years of evolution and injection of star-formed elements.

At the high temperatures under which the Big Bang formed the first elements, the gas was entirely ionized, electrons stripped completely from atoms. In such a gas, as soon as an electron is captured by a proton to make an atom, it is immediately ripped away by absorbing an energetic photon. Matter and radiation are therefore tightly coupled. A hundred thousand years after the first element formation, the gas had thinned and cooled to the point (about 7000 kelvin) that electrons could combine with nuclei to make permanent neutral atoms. Radiation then no longer interacted strongly with matter. As a result, radiation could escape the gas and run free.

Big Bang theory was developed in the 1940s principally by George Gamov and his group of researchers. At the time, they predicted that this free-roaming radiation should be detectable. These photons would be trapped into the fabric of space. Expansion of space over some 10 to 15 billion years would have stretched them out, would have reddened and "cooled" them to under 10 degrees kelvin. In 1965 two engineers, working at Bell Labs, found space to be permeated by background radio radiation produced at only three degrees kelvin, a discovery that even more powerfully supports the Big Bang.

This "cosmic background radiation" is not perfectly uniform, but is filled with fluctuations that seem to reveal similar fluctuations in the distribution of mass – including what is now called dark matter – that later became galaxies and clusters of galaxies. When, with the Hubble Space Telescope, we look at distant galaxies from which light took so long to come, we are also looking back into time, though not so far as we do with the cosmic background radiation. Light from a galaxy a billion

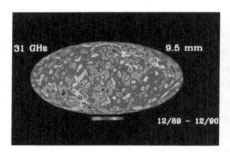

Figure 9.11. (See also Plate XXVIII.) This whole-sky map shows the brightness of the cosmic background radiation, which shines at a frigid three degrees above absolute zero, as observed by the *COBE* (= Cosmic Background Explorer) satellite. Tiny fluctuations of only a millionth of a degree reflect variations in density from which galaxies and their clusters ultimately arose. [NASA, G. F. Smoot *et al.*]

Figure 9.12. (See also Plate XXIX.) The Hubble Deep Field, a 100-hour exposure with the Hubble Space Telescope, reveals galaxies of all kinds terribly far away, to billions of light-years, allowing us to look billions of years back into the past. [STScI and NASA.]

light-years away took a billion years to get here, so we see the galaxy as it was a billion years ago. The Universe therefore itself presents us with a way of seeing it not as it is, but as it was. We see that galaxies were different back then, that the earliest systems were smaller pieces that accumulated to form the larger galaxies of today, and that even our own may have been the product of mergers, consistent with what our local system is telling us. We have now come full circle, and again can look at our own Galaxy and at the oldest stars of all.

## Return to the Galaxy

If Big Bang theory is correct – and all lines of evidence suggest that it is – the initial contracting cloud of matter that made the Galaxy and that generated the globular clusters should have been nearly all hydrogen and helium. There should have been no metals at all except a tiny amount of lithium. Yet even the most metal-deficient globular cluster has a significant metal content, even if only somewhat more than a few-thousandths that found in the Sun (at that level about one iron atom for every 10 million hydrogen atoms).

Globular clusters, however, are not the only constituents of the halo. There are a vast number of free-running stars that are not allied with clusters. Perhaps we can find the zero-metal "Population III" stars among them. All pursuits of such stars have been fruitless. Astronomers have found many that indeed have fewer metals. The current record – about a ten-thousandth the solar iron content – belongs to faint, twelfth magnitude CD $-38°245$ (a star in the nineteenth century's renowned "Cordoba Survey") in the southern constellation Sculptor. From the theory, CD $-38°245$ must be one of the oldest stars of all. But nowhere can we find the *very* oldest stars that should exist, nowhere do we find zero iron. Is there something wrong with the Big Bang after all? Have we so far not just found such stars? Or do such stars no longer exist?

Relief from scientific anxiety is found in the patterns of chemical compositions. As we go to older stars, those for which the iron abundance is below about a tenth the solar value, oxygen and other even-numbered elements (including magnesium, silicon, sulfur, etc.) become overabundant with respect to iron. There are fewer of these elements than in the Sun, but their deficiencies are just not as great as that for iron. A large fraction of the iron in the Universe is thought to come from the Type Ia, white dwarf, kind of supernovae. The lighter elements, however, the even–numbered ones beginning with oxygen, are in ratios expected for the Type II core-collapse variety, as are those of the observed r-process elements.

Astronomers therefore hypothesize that the first generation, zero–metal Population III, contained a set of massive stars that quickly exploded and seeded the early Galaxy with these initial elements. The iron content produced by a single small generation of core-collapse supernovae should produce a minimum iron abundance in the most metal-poor stars of about a ten-thousandth that in the Sun, and that is just what we find. The massive stars of the first generation are now gone, exploded

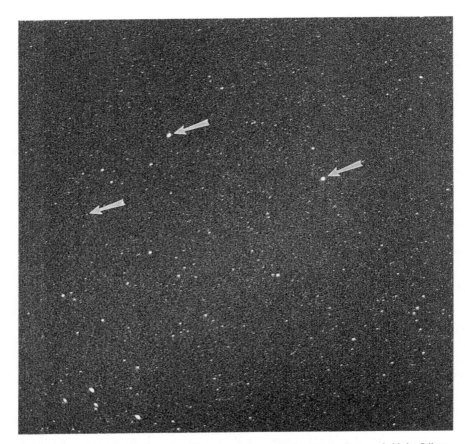

Figure 9.13. The right-hand and middle arrows, respectively point to Beta and Alpha Librae. Tucked into the dim constellation is the halo star HD 140283 (*left-hand arrow*). Visible in binoculars, HD 140283 has an iron abundance somewhat less than that of the most metal-poor globular cluster, which is still 20 times that of the most metal-poor star known. When you look at it in a darkened sky, you are looking at one of the oldest single stars in the Galaxy and are looking back to a time shortly after the Galaxy began. [Author's photograph.]

away, and if there are any low-mass counterparts they are simply so rare and faint, lost within the great Galactic halo, that it would be all but impossible to find one. Indeed, it will be extremely difficult to find any stars that are more metal-poor than the ones we now see.

White-dwarf supernovae, which create a third to over one-half the Galaxy's iron, had to wait for stellar evolution to make the first generation of white dwarfs, which takes a longer period of time than it does to make the core-collapse kind of supernovae, as the stars that make white dwarfs are less massive. Once white dwarfs could explode as supernovae, the iron abundance of newer generations of stars increased faster than the oxygen abundance, and the ratio of oxygen to iron declined, even as the absolute ratios of oxygen and iron to hydrogen rose. Inhomogeneities in abundances clearly seen among the oldest stars result from spotty evolution and a lack of uniform mixing in the early Galaxy.

We still know little about the details of star formation and how the metal content of the interstellar medium affects their formation and evolution. As a result we have little idea of what the initial distribution of stellar masses might have been. Neither do we know how these first stars were born without any Galactic dust or without prior supernovae to compress the interstellar medium. With no hot stars to heat the interstellar gases, cooling dust may not have been needed. One theory suggests that massive stars may have formed in the turbulence of the Big Bang even before the galaxies were created. The new galaxies would have been pre-seeded with metals, so there would be no zero-metal stars to find no matter how hard we look.

The remaining problems show the interrelations among the branches of the sciences, as cosmology and stellar theory involve particle physics, the physics of gravity, and a variety of other subjects. The oldest stars therefore provide a means with which to test the grand scheme of the Universe and to test our ideas of how stars and the Galaxy evolve. In our study of the oldest stars we stand at the forefront of a science in which we look back toward the distant unknown horizons of both space and time.

# Chapter 10

# The strangest stars

It is not the ordinary that drives scientific curiosity so much as the odd, the strange, the mysterious. And Nature accommodates with stars so weird as to be beyond imagination. We have walked the rim of the HR diagram, wandering a path that takes us through a forest filled with stellar extrema. By definition, the stars we have encountered are odd; they are, after all, the hottest and coolest, the brightest and dimmest, the largest and smallest, even the youngest and oldest.

What an astonishing array of stellar properties! Stars range from those that would fill much of the Solar System to those that could be placed in a small town: VV Cephei could hold $10^{24}$ Crab pulsars. The brightest stars have absolute visual magnitudes near $-10$, the dimmest (excluding the brown dwarfs that do not burn hydrogen) near 21. This range of 31 magnitudes corresponds to a visual intensity ratio of over two trillion, $2.5 \times 10^{12}$. RG 0050 would be visible to the naked eye only if it were closer than 0.006 light-years, or 4000 astronomical units; it would have to be within the Oort Comet Cloud, that is, actually inside the confines of our own extended Solar System. Cygnus OB2 #12, however, could be seen without aid even if over 50,000 light-years away; if there were no intervening dust to dim its brilliant light (and there is a great deal) we could make it out even if it were half the way across the Galaxy. The hottest stars (discounting neutron stars) have temperatures over 100 times those of the coolest. Temperature extremes are so great that the stars become oddly dim because most of the radiation is produced in invisible-wavelength bands: cool stars like Mira generate most of their light in the long-wavelength infrared, whereas the central star of NGC 2440 releases over 90% of its photons in the high-energy ultraviolet.

Near the edge, stellar behavior runs amok, allowing an amazing variety of pathological symptoms. The coolest giants can alter their chemical compositions, changing from oxygen-rich to carbon-rich, becoming carbon stars like R Leporis. The low-temperature giants are so distended that they are evaporating before our eyes.

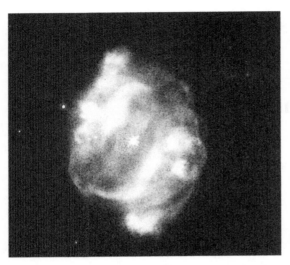

Figure 10.1. The planetary nebulae, represented here by the almost hopelessly complex NGC 5307, themselves represent the evolution of the intermediate-mass stars, those like the Sun. What was once among the larger and coolest of stars is transformed into one of the hottest stars and then into one of the smallest of stars, a white dwarf. [B. Balick (University of Washington), STScI, and NASA.]

Matter flows outward from the carbon star IRC + 10216 at such a rate (over $10^{-5}$ solar masses per year, a hundred million times that of our solar wind) that condensing dust hides the actual star from view.

The cool giants become even stranger when coupled to companions. Mass transferred from a lower main sequence star to a white dwarf can produce a nova or even a supernova; from a giant, flowing mass makes a symbiotic star that, like CH Cygni, can jump in brightness and flash like a celestial firefly. Companions to advanced giants may be responsible for the shapes of many planetary nebulae. Binaries among the largest stars are also responsible for amazingly long eclipses. It takes our Moon a mere three hours to complete a solar eclipse, and a transit of Venus (wherein the planet crosses in front of the Sun) lasts only a day. But the supergiant component of VV Cephei is so large that every two decades the orbiting O star is hidden for 1.2 years.

The brightest stars contain some of the oddest members of all, the "luminous blue variables" and the Wolf–Rayet stars. Among the LBVs we find weird P Cygni and Eta Carinae. Each is losing matter at an enormous rate, Eta as much as a thousandth of a solar mass per year, and both are very unstable. P Cygni, now fifth magnitude, brightened to first in the year 1600, and Eta Carinae, which is just barely visible to the naked eye, rivaled Canopus in 1844 and lost mass at over a tenth of a solar mass per year. Eta lost mass so fast after its last episode of brightening that it too – or perhaps a close companion – buried it within its own effluvia.

The brightest stars are also sometimes quite hot: the most luminous star in the Galaxy, HD 93129A, has a temperature around 50,000 K. But they are no match for the nuclei of the planetary nebulae, which are known to extend to 220,000 K. And as strange as the hottest stars are as a class they too have their own outstanding individuals. The central stars of a handful of planetary nebulae are surrounded by dense knots of nearly pure helium buried deep within the surrounding more-normal planetaries. In the process of dying as a white dwarf, the stars apparently suffered internal readjustments (sudden returns to helium burning) and re-expanded to giant proportions, allowing helium-rich layers to be lost.

From these superlatives, descend to stars with the least, the smallest, the white dwarfs, and the faintest, the red dwarfs. Sirius *B* rivals Earth in size, but has a mass comparable to the Sun and consequently (like all white dwarfs) a density that is quite beyond human comprehension. A cubic centimeter sampled from the interior would weigh a metric ton; and it is still a gas. Truly odd characters abound: stars with no surface helium or lacking in hydrogen, and a few with magnetic fields millions of times greater than that of our Earth.

And these pale beside neutron stars and pulsars, which are another hundred million times denser, with magnetic fields that produce polar beams of radio energy so intense that they would be lethal at the Earth's distance from the Sun, and can be "seen" – detected with radio telescopes – across a good portion of the Galaxy. The Crab Nebula pulsar, a star perhaps double the solar mass, is so energetic that it radiates not only in the optical, but produces X- and gamma rays as well. Such stars are hotter even than the central stars of planetary nebulae, with temperatures of perhaps a million kelvin. Double stars then rear their heads once more to create the millisecond pulsars that destroy their mates.

The dim red dwarfs have no such spectacular claim to fame, but they make up for it in population: they overwhelm anything else with their near-countless number, but are all so faint that none can be seen with the naked eye. Even these, as quiet and reclusive as they are, harbor surprises. Some, such as the closest star to the Earth, Proxima Centauri, produce shattering solar-like flares that involve the whole star and can brighten it ten-fold. The set also contains the mysterious brown dwarfs, failed submassive stars that never become hot enough to produce thermonuclear reactions, bodies that represent something of a bridge between stars and planets.

Even more remarkable than the properties of the individual groups of these extreme stars are the linkages between them as one kind transforms itself into another under the unstoppable forces of gravity and stellar evolution. The coolest stars – the class M asymptotic branch giants – emerge from their cocoons of dust to become the hottest stars when seen wrapped within their funeral wreaths of planetary nebulae. These then become the smallest of "normal" stars, the white dwarfs. The brightest stars, the blue O dwarfs, almost magically, bloom into the largest stars, the red M supergiants. In some form – perhaps as luminous blue variables, red or blue supergiants,

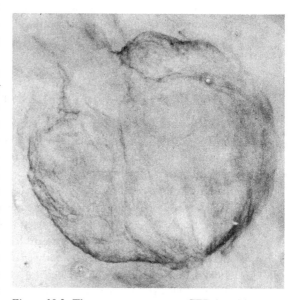

Figure 10.2. The supernova remnant CTB-1 and its shock wave (seen here as a negative image) expand into space, epitomizing the transformation of the brightest and largest stars into the very smallest. [R. A. Fesen *et al.*, Case Western Reserve University Burrell Schmidt.]

213

or the hot blue Wolf–Rayet stars – these brilliant bodies are destined to explode and produce the smallest of all stellar bodies, the neutron stars, or at the most extreme limit of all, the ultimate in stellar evolution, black holes, into which the stars disappear into punctures of spacetime. The transformations are wonderfully symmetrical. The bigger the initial star, the smaller the diameter of the remnant after evolution is over, which given that evolution is run by gravity, makes complete sense.

And all of the extremes, from biggest to smallest, coolest to hottest, dimmest to brightest, had to be born as well, so all passed through the phase of being "youngest stars," and among these, in addition to the classic T Tauri stars, are also the brightest stars that die so very quickly. The dimmest and coolest have the distinction of also being found among the oldest stars, from which we can explore the very age and history of our Galaxy, even portions of the age and history of our Universe.

Yet something is missing. As explorers we have wandered the shores of the continent of the HR diagram; have climbed the heights of its mountains, descended into its chasms, and have visited its volcanoes. But we have ignored the great plains of the heartland. That is where we find the Sun and the quiet well-behaved stars of the main sequence, so if we are searching for superlatives, why look there? But as in central Africa we come across the Great Rift, and in interior Asia towering Everest, surprises also await in the vast midlands of the HR countryside.

So in our travels, while we still trek in familiar territory, at the edges of the HR diagram to look at some of the odder members of the extreme stars so far examined, we also turn away from these outer fringes and sample some of the additional wonders to be found within the depths of the interior, as we look at just a sample of "the strangest stars."

## OH 231.8 + 4.2: The Calabash Nebula

OH/IR stars are cool Miras – long-period variables – that are buried within their dusty winds. They are oxygen-rich (that is, not carbon stars), glow brightly in the infrared from their heated dust, and contain OH and other molecules that act as natural masers. A considerable number are known. One is outstandingly odd as it is placed within a weird-looking gourd-shaped nebula known as the "Calabash." Allied with the well-known cluster M46, the object is some 4000 light-years distant and huge, nearly 1.5 light-years from one end to the other. The smaller northern end is tipped forward, and is moving outward from the center at a speed of 110 kilometers per second, whereas the larger southern end is moving away almost twice as fast, at 190 km/s. From the size and velocity of expansion we estimate the Calabash's age at about 1400 years. At the lower end are three condensations that look like (and indeed are) old-star versions of Herbig–Haro objects. Buried in the middle is a luminous, very cool long-period Mira variable (called QX Puppis) that has an M9 spectral class and a very long 670-day cycle of oscillation. The star is optically obscured by dust; we know of its nature from the infrared and because its light is scattered from the dust that surrounds it. It probably began life as a three solar-mass main sequence B star, and is now in the process of whittling itself down.

The bubbles are gas flows from the buried Mira that appear to be highly directed by a very dense ring of dust. All advanced giant stars lose matter at a fierce rate, as high as a ten-thousandth of a solar mass per year. In this unusual case only the ejected gas that is directed along the disk's polar axis can escape. It does so at such a speed that huge outer bubbles and the Herbig–Haro lookalikes are created by shock waves that develop when the matter rams against the ambient medium, which is most likely atmospheric material that was driven away in prior episodes of mass loss. In death, we see a reprise of behavior much like that we would have witnessed during the star's birthing process.

The next step is almost certainly the birth of a planetary nebula, which will develop when the star loses its remaining outer envelope, the fast wind begins to shovel the old wind, and the core becomes hot enough to ionize its own ejecta. The currently-visible bubbles are much too far away from the star to be an actual planetary; the new one will develop down in the hidden central region. The outflow pattern will probably be repeated, however, and sometime in the astronomically near future we will see a bipolar (twin-lobed) planetary nebula slowly develop along this same axis around the fiery nuclear-burning core that began its optically visible life as a youthful T Tauri star. By then the Calabash itself will have nearly vanished into the depths of space, perhaps to be seen only as a faint outer halo that surrounds the inner planetary.

Most "protoplanetary nebulae," reflection nebulae that are destined to be planetaries, have warmer central stars, stripped AGB cores that are heating through classes F or A. A class M central core is rare, indeed unique, suggesting that we have caught the developing nebula in a very brief stage of its life, the star thus providing us with a powerful link between the coolest and hottest stars.

Figure 10.3. (See also Plate XXX.) The Calabash Nebula, named after its odd gourd-like shape, surrounds a long-period M9 Mira that lies at the dense neck of the object where the two bubbles meet. The star is optically hidden from view by a thick disk of dust; gas streams out through the poles producing shock waves as it batters the surrounding medium. [B. Reipurth, European Southern Observatory.]

## FG Sagittae

The planetary nebulae that develop from such Miras are really rather well understood. The lovely structures, produced by the last stages of mass loss, will blend with the interstellar gas, leaving behind a naked white dwarf. A number, however, present us with puzzles. A very faint planetary nebula in the constellation

Figure 10.4. FG Sagittae, centered in an otherwise ordinary planetary nebula, brightened from 14th to 10th magnitude between 1894 and 1965, changed its spectral class from B4 to K0 between 1960 and 1983, and is phenomenally rich in chemical elements heavier than iron. [Lick Observatory photograph from an article in the *Astrophysical Journal*, 1968, by G. H. Herbig and A. A. Boyarchuk.]

Sagitta is renowned for its central star. The star was first recorded in 1894 with a magnitude (measured in blue light) of 13.6. Over the past century it has brightened steadily and is now up to tenth magnitude. As a result, it was assigned the variable-star name FG Sagittae. In 1955, Karl Henize (who later became a Shuttle astronaut) discovered the nebula and catalogued it as He1-5, though the nebula is loosely known by the star's name as well.

In itself, the variation of the star is not all that strange, as many stars exhibit large, long-term changes: Eta Carinae and other luminous blue variables, "slow novae" such as RR Telescopii, the FU Orionis stars, and even a number of symbiotic stars. But this behavior is outstandingly odd for the central star of a planetary nebula. Moreover, the brightness change has been accompanied by unique spectrum variations. In 1955, the star was classified as a B4 supergiant. Even that is not all that unusual as several planetaries have apparently unorthodox stars at their centers that are clearly brighter binary companions to the true nuclei. Further observation, however, showed that the star was rapidly changing spectral class and cooling by about 300 degrees kelvin per year; by 1983 it had dropped to a temperature that took it into the G star range, where it remained a brilliant supergiant.

Stranger still are the kinds of spectrum lines that the star developed. Ordinary stars exhibit numerous spectral features caused by moderately light metals such as iron, chromium, titanium, and vanadium. FG Sagittae, though, began to exhibit spectrum lines belonging to such odd elements as barium, zirconium, yttrium and several rare earths. The rare earths as a group fall between barium, atomic number 56, and hafnium, number 71. They are over twice as heavy as iron, and include such under-appreciated elements as cerium, praseodymium, neodymium, promethium, samarium, and gadolinium. As their collective name implies, they are rare. Within the Sun there is one atom of cerium, the most common, for every 28 billion hydrogen atoms and every 1.3 million iron atoms.

Within the star of FG Sagittae these rare elements are not just faintly present, but dominate the spectrum. Yttrium and barium are enriched well beyond 30 times normal. On the other hand, the common "iron-peak" elements (those near this most common of metals), which include nickel, have nearly faded away. The elements that make the observed absorption lines can only be produced by the "s-process," in which heavier elements are created by the slow capture of neutrons onto lighter nuclei. Many advanced red giants are rich in s-process elements, the mechanism producing the zirconium of S stars and even radioactive technetium in red giants. These elements are created deep within the nuclear-burning hearts of the stars and brought to the surface by convection. To see such behavior in the nucleus of a planetary nebula is bizarre, indicating some kind of odd advanced evolutionary status. By its very nature, since any such evolutionary condition is rare and fleeting, this behavior strongly indicates that the star is indeed the true nucleus of the planetary and not just a binary companion.

Odder still is the increase of elements like europium and gadolinium, which are usually attributed to the r-process, rapid neutron capture that seems only to happen in supernovae. A small portion of these elements can in fact be made by the s-process, and the star must be working overtime inside to make them so prominent.

Whatever is happening to produce such changes and rampant alchemy is not really known. The star has shown P Cygni lines that indicate it to be losing mass. The spectral changes seem actually to be caused by a cool expanding layer overlying the true star within: much the same phenomenon that produces spectral changes in the luminous blue variables. The star may in fact be caught in the act of re-birthing itself as a giant star, having fired up its internal helium layer as a not-quite-dead white dwarf. Convection then brings the new stuff, made in hundreds of nuclear reactions, to the top. Because of the low surface gravity, the star is now producing another planetary nebula inside the first one, and may someday be seen as something like Abell 78 and Abell 30, which were explored among the hottest stars.

FG Sagittae (with its nebula) is at the other end of the evolutionary trail from the Calabash, not a planetary at the beginning of its life, but at the end. Moreover, FG Sagittae can quickly dim by many magnitudes as its wind sporadically pumps carbon-rich matter that condenses into obscuring carbon dust. It may be in the process of becoming an R Coronae Borealis star, to which we now turn.

## R Coronae Borealis

Dip now into the HR diagram's midlands. A naked-eye star in the curve of Corona Borealis, the Northern Crown, is avidly watched by dedicated amateur astronomers every clear night that the little constellation is visible in hopes of being the first to watch it disappear. Though the star's spectrum is odd, it is usually classified as a G-type supergiant. Unlike most G supergiants, however, the star will without warning drop in brightness by as much as eight magnitudes, making it impossible to see without at least a 20-centimeter telescope. Complete recovery back to sixth magnitude can take well over a year. It has been called an "inverse nova," its strange light-

Figure 10.5. R Coronae Borealis (*arrow*) is easily seen with the naked eye within the curve of the Northern Crown. [Author's photograph.]

curve looking vaguely like that of a common exploding star turned upside-down. There is, of course, no relation between R Coronae Borealis and a nova. The sudden plunge is caused by the veiling of the star by carbon dust – soot – of its own making during the onset of a sudden episode of mass loss. There appears to be no periodicity to the events, although there is an average interval of three to four years separating them; prediction, however, is quite impossible.

R Coronae Borealis – discovered in 1795 – is the prototype of a class of about thirty G- to A-type giants and supergiants. Other particularly well-known ones are 6.5 magnitude RY Sagittarii, tenth magnitude SU Tauri, and the equally faint southern star S Apodis (in the obscure constellation Apus). All behave quite similarly. Their spectra show them to be highly deficient in hydrogen, which is normally the most common element found in stars. Instead, helium dominates, and carbon is greatly enriched, consistent with the periodic veiling of the stars by carbon dust. During the brightness decline, the spectra develop emission lines characteristic of a chromosphere (the layer immediately surrounding the bright surface of the Sun). The dust therefore apparently forms relatively close to the surface, hiding the bright surface of the star but allowing us to see its extended outer regions.

R Coronae Borealis variables show evidence for pulsation over a period of a few tens of days, which can be explained by a Cepheid-like instability strip. One possible, but contended, explanation is that the pulsations drive off puffs of gas that produce dust fairly close to the stellar surface. If the puff happens to lie between it and the Earth, the star will fade, an explanation that accounts for the randomness of the event. As a result of many such episodes, the whole star is enmeshed in a great warm dusty cloud of its own making, one that was easily seen in the infrared by the orbiting infrared observatory satellite *IRAS*. No one knows why mass loss should be so chaotic.

The stars are highly evolved, though just where they are coming from and where they are going is at present not at all clear. From observation of those in the

Magellanic Clouds, they appear to have absolute magnitudes of perhaps −5, giving them bright giant or even supergiant status. Further evidence for a high degree of evolution comes from the dominance of helium. These odd variables must have gone through great earlier episodes of mass loss to have dissipated their initial hydrogen envelopes, and we believe them to be in a post-asymptotic giant branch phase. Some astronomers have suggested that they are in the process of becoming planetary nebulae. Others have proposed that they already have *been* planetary nebula central stars (like FG Sagittae), but have undergone changes that made them into giant status. These objects remain an enduring enigma: one that can be watched from your own back yard.

## Epsilon Aurigae

The oddities to be found among single stars pale when compared to those encountered among binaries. We have met strange doubles many times: novae that explode when dwarfs dump matter onto white dwarfs; symbiotics in which hot white dwarfs receive sprays of matter from cool giants; solar system-sized VV Cephei, which hides its O-star companion for over a year; X-ray binaries with neutron-star components; millisecond pulsars that devour their mates; even stars that seem to harbor not stellar companions but planets.

Near Auriga's Capella, the backyard observer sees a familiar triangle, the "Kids" of the celestial she-goat. Two of the stars are variable. Zeta, the faintest, is a lesser version of VV Cephei, in which a B5 dwarf disappears behind a bright K4 giant for 38 days during the 2.6-year binary period, causing a near-imperceptible drop of 0.2 magnitudes.

The show really belongs to the northern one of the trio, Epsilon. The primary is an F0 supergiant. Usually the giant (or supergiant) is the large body of an eclipsing system, as in VV Cephei and Zeta Aurigae, and the one that produces the major eclipse when it passes in front of a smaller bright companion. When the smaller star is forward, it cuts off only a portion of the light of the big one to produce a much

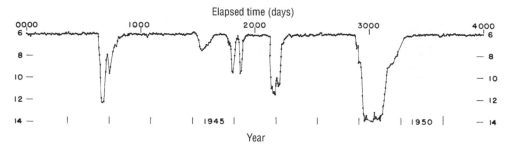

Figure 10.6. R Coronae Borealis's light-curve over a 4000-day interval shows it to dive suddenly from 6th apparent visual magnitude to 10th or fainter. A running day number is across the top. [AAVSO light-curve of R Coronae Borealis from 1941 to 1952. Courtesy of American Association of Variable Star Observers.]

shallower decline. But in Epsilon Aurigae the eclipse is produced every 27 years when part of the *supergiant's* light is cut off for an astonishing two years. The companion must therefore be even larger than a supergiant, yet no accompanying starlight is evident either during or outside the eclipse.

Infrared observations do show a glow from a body heated to 500 kelvin. It is believed to be a huge distended disk of dust orbiting the supergiant. The supergiant's spectrum, however, exhibits regular variations in radial velocity, showing that it is being significantly deflected by the dark companion, which indicates a mass far larger than any cloud of dust, however thick, could possibly have. One explanation – indeed the only one considered likely – is that a pair of B stars orbit closely around each other inside the dust cloud and by their gravity keep it from dissipating. (A single O star with enough mass to cause the supergiant's deflection would produce so much radiation that it ought to be visible.) The whole system in turn revolves about the supergiant. The origin of the dust is unknown, and so far as we know, the stellar trio is unique. We will probably have to wait until the next eclipse in the year 2009 to test these ideas and to find out the truth about the object.

## Chi Lupi and friends

In the middle of the HR diagram, among main sequence classes B, A, and F, are numerous stars with peculiar chemical compositions. The "metallic line stars," though low in some metals like calcium and scandium, are greatly enriched in others such as copper and zinc. They are typically of classes A and F (called Am and Fm).

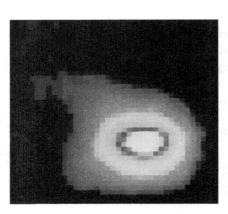

Figure 10.7. (See also Plate XXXI.) An infrared view of R Coronae Borealis shows a huge cloud of glowing dust several minutes of arc and over a light-year across that surrounds the star proper. The cloud is the dusty remains of many episodes of the mass loss that have produced the random drops in visual brightness. [AURA/NOAO/NSF.]

Relative to what is found in the Sun, the stars are also enriched in rare earths, the elements that form a side-branch of the periodic table from element number 57 through 71 and that are encountered in FG Sagittae. More dramatic are the "magnetic stars" (Ap and Fp, "p" for "peculiar"). These possess global magnetic fields (measured from the Zeeman effect, which splits spectrum lines) that are thousands – even tens of thousands – of times stronger than the solar field. Here, chromium and strontium can be enriched by a factor of 100 and the rare earths by factors of 1000 or more. The abundances are not even distributed evenly on the star, the spotty effect visible as rotation carries different regions in and out of view. Raising the oddness level, the "mercury–manganese

Figure 10.8. Auriga rises with bright Capella accompanied by her three "Kids." The northernmost (*arrow*) is the strange eclipser Epsilon Aurigae. Zeta Aurigae is the uppermost of the pair down and to the right of Epsilon. [Author's photograph.]

stars" (usually warmer class B) show elements like manganese and yttrium enhanced thousands of times and mercury enriched by factors of *tens* of thousands of times relative to solar abundances.

These oddities are represented by Chi Lupi, an easily visible fourth magnitude star that lies just south of the head of Scorpius in the southern Milky Way constellation Lupus, the Wolf. It is a binary whose components are so close that they cannot be separated and thus yield a composite spectrum. Here two categories come together, the brighter member of the pair a B9 mercury–manganese star, the other an A2 metallic line star. The pair is so outstanding in its properties that its study has been designated a "pathfinder" project for the Hubble Space Telescope. It is ideal for the examination of the abundances of very heavy elements including platinum, gold, and bismuth.

The origins of the these "chemically peculiar" stars were hotly debated for years. One popular theory held that they were contaminated through binary evolution. The more massive of a pair must evolve first. In its giant state it may bring

Figure 10.9. Chi Lupi (*lower arrow*), a prominent mercury–manganese star, shines brightly against the faint stars of the Milky Way near Scorpius. Antares (*upper arrow*), enmeshed in nebulosity, and its two flanking stars Sigma and Tau Scorpii, dominate at upper left. To the right of Antares is the bright globular cluster M4. [*Atlas of the Milky Way*, F. E. Ross and M. R. Calvert, University of Chicago Press, 1934. Copyright Part I 1934 by the University of Chicago. All rights reserved. Published June 1934.]

Extreme Stars

freshly-made elements to the surface. If the two stars are close enough, the giant may deposit some of its mass onto the remaining main sequence star, thereby enriching it. This process in fact produces "barium stars," giants enriched not only in the eponymous barium, but also in carbon and in s-process elements in general. They are all doubles with white-dwarf companions, and were almost certainly contaminated by their partners during their earlier evolution. However, neither of Chi Lupi's stars qualifies as being that evolved, and there is no evidence for a white dwarf. Another theory suggested contamination by accretion from the interstellar medium.

Astronomers finally come to the broad consensus that the strange abundances of the metallic, magnetic, and mercury–manganese stars, epitomized so well by Chi Lupi, are caused by diffusion, by physical segregation of the elements. The outer layers of the Sun roil with convection, which keeps the gases stirred up. Convection quiets in the hotter stars, however, the atmospheres of the A and B stars being quite still. Moreover, the chemically peculiar stars are slow rotators, rotation also acting to stir a star. Different atoms are thus free to drift according to whatever forces act upon them. Some, heavier ones, sink under the action of gravity, whereas others, pushed by the absorption of starlight, are lifted upward. As the elements become physically segregated, winds that depend on temperature blow some of them preferentially away, helping give the star its peculiar character and actually changing the average chemical composition. Magnetic fields add spice to the brew, acting on yet other atoms and separating some into "starspots." The processes are dramatic variations on the settling of helium that makes the DA white dwarfs. Full understanding of the mechanisms will lead to improved knowledge of the interactions between radiation, magnetic fields, and matter, and improved compositions of a variety of other kinds of stars in which the effects are present but not so severe.

## SS 433

Return to the edge of the HR diagram. Ten thousand light-years away, within a degree-wide supernova remnant called W 50, lies the 14th magnitude star SS 433, listed unobtrusively in a catalogue of emission-line stars compiled by C. B. Stephenson and N. Sanduleak. The first detailed spectrogram, published in 1978, showed common emissions of hydrogen and helium, but also some features at odd wavelengths that could not immediately be identified. Further investigation revealed that these lines were moving wildly through the spectrum, to a degree never before approached by any astronomical object. They were quickly identified again as hydrogen and helium, but with huge Doppler shifts that revealed velocities a good fraction that of light and that were variable over a 164-day period.

There are two sets of these lines, one shifted to shorter wavelengths (into the blue), indicating approaching gas, and the other toward longer, into the red, showing material flowing away from us. At maximum we can see a recession speed of 50,000 kilometers per second and an approach of 30,000. The variations are entirely regular with an average Doppler shift appropriate to a speed of 12,000 kilometers per second

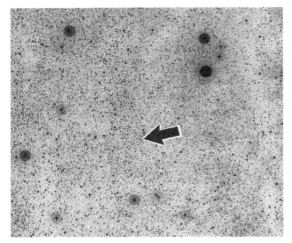

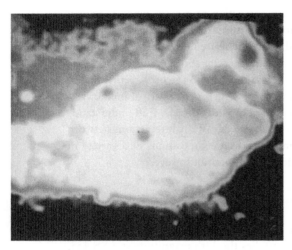

Figure 10.10. In this negative image, the star SS 433 (*top, arrow*) is centered in the supernova remnant W 50. The gaseous remnant (*bottom*) is quite bright in the radio part of the spectrum but is only faintly visible in the optical. [*Top*: Palomar Observatory, California Institute of Technology, S. van den Bergh; *bottom*: VLA/NRAO/AUI, J. R. Dickel.]

in recession and with the blue and red sets periodically crossing over one another.

Velocities like these are seen in receding galaxies, and it would at first appear either that this must be some kind of strange extragalactic source, or that it has to be sailing quickly out of the Galaxy. However the common stationary set of lines shows a radial velocity of only 70 kilometers per second, not at all unusual among stars. That and the coincidence with a supernova remnant shows us that something else, something very strange indeed, must be going on.

The stationary lines actually have slightly variable Doppler shifts over a 13-day period, revealing that the star we see has an invisible binary companion. The star appears to be throwing off opposing rotating jets as if it were some wildly accelerating celestial lawn sprinkler. We can account for them with matter that is flowing from an O- or B-type supergiant into an accretion disk around a collapsed stellar object. The characteristics of the Doppler shifts induced by binary action are such that the small companion is likely to be a neutron star, although a black hole cannot be ruled out. We thus have the by-now familiar picture of columns of gas that stream back out of the star along the poles of the disk, much as we see for T Tauri stars and the Calabash, but with enormously more violence as befits the energetics and powerful gravitational field of the collapsed remnant of a supernova.

But what causes the variations in velocity? They can be explained if the disk is tipped relative to the equatorial plane of the binary orbit. There is a clear analogy with the Earth and its orbit. Our planet's rotational axis is tilted with respect to its orbital plane through the well-known angle of 23.5 degrees, which gives us our changes in seasons. As a result of the lunar and solar gravitational pulls on the ter-

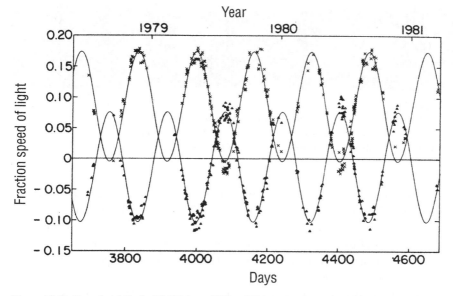

Figure 10.11. Doppler shifts in SS 433 from 1978 to 1982 (in fractions of the speed of light plotted against a running count of days) show a 164-day period in emission-line positions, hence velocities. The observed velocities for the two sets of components cross over one another. [From an article by B. Margon that appeared in *Annual Review of Astronomy and Astrophysics*, 1984.]

restrial equatorial bulge (caused by our rotation), the Earth's axis wobbles – precesses – over a 26,000-year period. The same thing is happening in SS 433, but over a vastly shorter interval. The gravity of the large component causes the accretion disk that surrounds the neutron star (or black hole) to precess with a 164-day period. What we see can be explained by an accretion disk tilted to the line of sight, from which jets gush outward in the perpendicular direction at 26 percent the speed of light.

What, however, about the huge apparent 12,000 km/s radial velocity that seems to be taking the star out of the Galaxy? It is the result of the enormous speed of the jet material as it pours out of the deep gravitational well of the collapsed object. Since the matter is moving at a good fraction the speed of light, we must include the effects of Einstein's relativity, which predicts a "second-order Doppler effect" that perfectly fits the observations. Not only is SS 433 a fascinating object in its own right, but it provides yet another proof of this famous theory – as if by now any is actually needed.

Awful violence of this sort ought to have an effect on the immediate environment of the star, and indeed it does. Radio observations show corkscrewing gaseous streams a few seconds of arc across coming out of the star, as would be expected from the swinging beams of gas. These are aligned in space with a bar of X-ray emission nearly a degree long. The jets seem to have an effect over 100 light-years away from the central source! Moreover, both of these are aligned with the major axis of the

Figure 10.12. Over a period of 10 days (*left to right*), blobs of matter move outward from the center of GRS 1915 + 105 at speeds that seem to exceed that of light. The actual speed, while less than that of light, is likely to be in excess of 270,000 km/s (0.92 that of light). The geometry of the object merely fools us into seeing super-light speeds. [*MERLIN*, R. Fender *et al.*]

supernova remnant, as revealed through radio imaging; they seem to be largely responsible for the bulged shaped of the object. The star and W 50, which is nearly 200 light-years across, are now sweeping up 30,000 solar masses of surrounding interstellar gas, further testimony to the immense energies involved.

This amazing object, apparently created in a great supernova blast perhaps 5000 or so years ago, is – yet more amazingly – not the most extreme. Other, similar, objects have flows of gas that move even faster. They were found initially with the Compton Gamma Ray satellite as sources of powerful X-ray and gamma-ray emission. Radio observations subsequently revealed gas blobs moving outward at speeds well exceeding that of light! Such behavior of course is not physically possible.

The key is that we see similar flows coming from the ultraluminous cores of distant galaxies known as quasars. In these, the flows are interpreted as jets ejected from massive central black holes. The jets are moving close to the speed of light in a direction more-or-less toward us. The geometry of the situation (the tilt of the flows to the line of sight), coupled with the finite speed of light, tricks us into seeing blobs of gas moving faster than light when they are really not. Nevertheless, the similar flows that come from stars within our Galaxy exceed even the speeds of those from SS 433, and are estimated to move at speeds about 90 percent that of light. They are almost certainly also linked to supernovae, in that mass from a binary companion is being accelerated outward by a neutron star, or even by a black hole, all these phenomena demonstrating that there is a great deal yet to be learned about the phenomena of these devastating blasts.

## And finally

This chapter lists but a summary view, a sparse sample, a flavor, of the strange delights to be found amidst the normal stars of the nighttime sky. Anyone else writing it would certainly have selected others. No mention has been made, for example, of the contact binaries, the W Ursae Majoris stars, in which the surfaces of the whirling individuals actually touch each other; or of the FK Comae Berenices stars that are covered with dark starspots; of the RS Canum Venaticorum stars, synchronously rotating binaries (rotation coupled to orbital revolution) that are also covered with starspots; or of the madly spinning emission-line B stars that have ejected thick equatorial disks.

No star, however, seems to be quite so strange and intriguing as a seemingly ordinary dwarf, one of billions within our Galaxy, one that has no characteristics that an astronomer on a planet orbiting another star – at least one with our capabilities – could be aware of. Look then to our own Sun and to its treasure: Earth. A common description of Earth includes its insignificance as a tiny speck of matter that orbits an ordinary main sequence star on the fringe of just another spiral galaxy, one of two that dominates a poor cluster of galaxies, all on the fringe of a more massive cluster from which we are slowly moving away.

Look, however, at the stuff from which the Earth and the other small planets of the Solar System are made. By far the most common element in the Universe is hydrogen, followed by helium. Over 90% of the solar atoms are hydrogen atoms, helium constituting somewhat under 10%. All the rest account for 0.015% of the atoms, or (since they are heavier) about 1% of the total mass. Our Earth, however, has a pathetically small amount of hydrogen. It may seem like a lot to someone at the ocean shore, or to an astronaut looking at our planet from a distance and seeing the blue of the $H_2O$ seas. But the oceans average only about 3 kilometers deep on a planet 13,000 kilometers wide. They are no more than a thin film on the planet. There is even less helium, nearly all of it a by-product of radioactive decay within our planetary rocks.

In contrast to the Sun, the elements that make Earth are the heavier ones, from carbon on up in atomic number. The Earth's core, which holds about a third of the terrestrial mass, is 90% iron and 10% or so nickel. The outer mantle that makes almost all of the remaining 70% consists mostly of silicon and oxygen in the form of silicate rock – with a lot of other stuff, various metals, thrown in. The Earth is the solidified distillate of the solar nebula from which the Sun and planets were born 4.5 billion years ago. Almost all the abundant light matter, the hydrogen and helium, has

Figure 10.13. (See also Plate XXXII.) Home, our solid Earth, made mostly of the debris of supernovae. [NASA.]

Figure 10.14. (See also Plate XXXIII.) The grandeur of a living rose places the Sun among the true wonders of the Universe. [Author's photograph.]

been driven away, more accurately was not allowed to accumulate because of the high temperatures found near the forming Sun.

And where does this heavy stuff come from? From evolving stars, from giants, planetary nebulae, and most importantly from supernovae. Supernovae of both types are – from all our observations – the sources of all the iron in the Universe, all the silicon, probably all the oxygen, and of a huge proportion of the other heavy materials. If you assume about one supernova every 50 years over the 10-billion-year lifetime of the Galaxy's disk, 200 million of them have popped off. If the Earth, whose mass is $6 \times 10^{27}$ grams, is made primarily of supernova debris, and if supernovae contributed uniformly to the Earth (which they certainly did not), then each supernova conferred a mass to us of about $10^{19}$ grams (10 trillion metric tons): about the mass of a good-sized mountain, making the linkage between Earth and cosmos far more real. As the harbor to Earth – our planet now seen as the distillate not just of the solar nebula but of the Galaxy's supernovae – the Sun rightly takes a place among the strangest stars, at least as far as our present knowledge is concerned, no other "earths" (excluding the oddball pulsar planets) known.

What makes the Sun truly unusual though is not the nature of the Earth itself but the nature of what is on the Earth. The Sun shelters not just supernova debris but life, indeed intelligent life, perhaps the only such life that there is. Maybe in our search for stellar limits, for the grandest of stars in any category, we should give first rank to an otherwise ordinary G2 dwarf, one whose third planet has both distilled the power of supernovae and developed the life that allows the appreciation and comprehension not only of the remarkable stars presented here but of the entire Universe.

CPSIA information can be obtained at www.ICGtesting.com
Printed in the USA
LVOW03s0538150414

381679LV00004B/79/P